BOOK LOAN

Please return or renew it no later
than the last date shown below

SYSTEMS IN ORGANIZATIONS

BUGS + FEATURES

M. LYNNE MARKUS

Pitman

Boston · London · Melbourne · Toronto

Pitman Publishing Inc.
1020 Plain Street
Marshfield, Massachusetts 02050
Pitman Publishing Limited
128 Long Acre
London WC2E 9AN

Associated Companies
Pitman Publishing Pty. Ltd., Melbourne
Pitman Publishing New Zealand Ltd., Wellington
Copp Clark Pitman, Toronto

Library of Congress Cataloging in Publication Data
Markus, M. Lynne.
    Systems in organizations.

    Bibliography: p. 228
    Includes index.
    1. Organization.   2. System analysis.   I. Title.
HD31.M2983   1984   658.4'06   83-22121
ISBN 0-273-01799-3

Manufactured in the United States of America
10 9 8 7 6 5 4 3 2 1

Contents

Preface

One of the most remarkable characteristics of systems is that organizational changes—like better performance—frequently accompany them. This characteristic makes building systems a useful tool for improving results. Unfortunately, however, the changes are not always for the better, and more often than not, the performance of departments is worse after a system is installed than before.

The natural tendency when observing these changes for the worse is to conclude that the system was somehow badly designed and to put one's hopes for better future systems into the training of system designers. However, not all changes for the worse can be attributed to the failure of a professional designer to translate accurately the idea for what a system should do into a system that does it. Sometimes the changes occur in areas of the organization that were never even considered when the system was designed. Occasionally it turns out that the system performs exactly as intended, and therefore, changes for the worse can only be blamed on the system.

This suggests the need for a much larger view of the system design problem than one that looks only at the activities of professional system designers. For the most part, managers in organizations, not system professionals, determine what systems are supposed to do, and they are certainly in a better position to anticipate unfortunate system side effects.

This book examines the entire process of designing systems within the organizations that adopt them—not only the activities of system professionals but also the roles of managers, external vendors and consultants. It addresses the processes of deciding what a system should do, of building it to do so and of integrating it into the organization.

Systems in Organizations: Bugs and Features presents a framework for identifying and explaining, predicting and controlling, the changes caused by systems. This framework is based on the simple but powerful theory, drawn from the social and behavioral sciences, that the impacts of systems are not caused by system technology or by the people and organizations that use them but by the interaction between specific system design features and the related features of the system's organizational setting. The book is intended for graduate-level students of management and information systems and for practicing managers who want to know what they can do about the systems that affect their performance as well as the people with whom they work.

PLAN OF THE BOOK

Chapter 1 examines one of the major obstacles to improving systems in organizations: lack of agreement among system designers and users about what the real or important problems are. Different views of the problems lead to different perspectives on their causes and different recommendations for action. Chapter 1 introduces the perspective on system problems that is used throughout the book.

Chapter 2 identifies five types of systems and describes their characteristic design features. This scheme is useful for classifying systems and identifying their impacts whether the systems are manual, computer based or based on a technology other than computers.

Chapter 3 describes the changes, both negative and positive, that may result when systems with certain design features are introduced into organizational settings with certain features. The chapter also explains how to perform an impact assessment for a specific system.

Chapter 4 describes and illustrates how impacts happen. It demonstrates that designers' intentions, however loosely translated into a real system, are fundamentally related to people's resistance to systems and to organizational changes for better or worse.

The remaining chapters in the book explore the problems that arise after someone has decided what the system will do. Chapter 5 describes the traditional methods system professionals use to build systems and some recent variations on them.

Chapter 6 discusses one important aspect of the organizational context of system building: the relationships between the system-using organization and the many external vendors of technology, systems, ideas and advice. The chapter shows how the dynamics of user/vendor interactions can influence both the system and eventual performance.

Chapter 7 examines a second aspect of the organizational context of system building: the way in which the system-building process is structured and managed inside the firms that use systems. The chapter explores the influence of the infrastructure on system success and organizational change.

Chapter 8 presents the practical recommendations derived from the interaction perspective for the effective design, implementation and support of systems in organizations.

Acknowledgments

I first became interested in systems as a doctoral student in organizational behavior at Case Western Reserve University. At first, I was only taking a hint from one of my advisors, who suggested that I would do well to pick a dissertation topic that capitalized on my undergraduate education in industrial engineering. After a brief introduction to systems, I was convinced that this was the area for me. Here, it seemed to me, was a vast uncharted territory for people with an orientation toward social and organizational studies. It seems that way to me still.

At Case, I learned about systems from the user's point of view. I wrote and rewrote several of the cases that I use as examples in the book—FIS, JHM, and Masada—and that helped me generate the perspective on systems presented here. From Case, I went to MIT where I was introduced to an entirely different viewpoint—that of the designer. I learned, in detail, the limitations of the technology and the ways in which these limitations are translated into constraints on the user. I learned about the computer world and the computing infrastructure. The intellectual atmosphere there challenged me to use my qualitative evidence to support crisp arguments. Later, at Arthur D. Little, Inc., I had the opportunity again and again to see the need for, and usefulness of, the ways of thinking about systems and acting on them described in the book.

I owe thanks to many people, more than I can name, for their intellectual guidance, constructive criticism or warm support. A few must be singled out. My dissertation committee at Case—Dave Brown, Jeff Hoffer,

Eric Neilsen and Bill Pasmore—struck the right balance between direction and approval. Rob Kling, at the University of California at Irvine, became my mentor in spite of the miles and has provided inspiration, counsel and friendship ever since. At MIT, four people stand out: Jack Rockart, Stu Madnick, Lotte Bailyn and Dick Beckhard. Jeff Pfeffer, at Stanford, gave me the incentive to write this book and has urged me on with his unfailing "Is it done yet?" Dan Robey, at Florida International University, became a wonderful colleague and friend and reviewed and commented on the manuscript in its multiple incarnations. Bill Roberts has been the type of publisher every first-time author hopes to have, reading and commenting along with the reviewers and becoming a good friend. John Aram, at Case Western Reserve University, provided many helpful and encouraging suggestions in his reviews of the book. Two anonymous reviewers offered valuable criticism. Students at several schools have challenged my ideas and my presentation of them and have provided advice, case material and encouragement. Colleagues at Arthur D. Little have supported and encouraged me while I finished the book at night and on weekends. Finally, the secretarial support staff at three institutions, Case Western Reserve University, the Sloan School of Management at MIT and Arthur D. Little, have been invaluable. (My office automation colleagues are speechless when I respond to their question, "What system did you write your book with?" by holding up my Montblanc fountain pen. The real system, however, was the secretarial support staff.)

In addition to those named, I owe thanks to many people whose contributions have been less direct but just as real. My parents and sisters have provided unfailing love and encouragement. Several dear personal friends have given me emotional and every other kind of support when I needed it and even when I didn't. I wish I could name all of you here. I especially thank Charles Donald Scales who, for all his loving support, has every right to refer to this as "our first book."

1

Bugs and Features

INTRODUCTION

Is that a bug or a feature? The question has the status of an in joke among the management information systems faculty and doctoral students at the Sloan School of Management at MIT. I heard the question for the first time on almost my first day there; I had not been there many weeks before I learned its origins.

The episode took place in about 1964 or 1965 at MIT's Laboratory for Computer Science. During the mid-1960s, certain computer vendors would occasionally give the lab a copy of software programs for evaluation before releasing them as products to customers. One such program was a Fortran compiler that was designed to translate programs written in the Fortran language into something the computer could compute. The evaluation revealed a so-called bug in the software.

The term *bug* usually refers to an error in the software instructions that tell the computer what to do. Because computers execute their instructions literally, errors in instructions produce results other than those the programmer intended. Bugs reputedly got their name because a short circuit in an early computer was discovered to have been caused by a moth.

The bug in the Fortran compiler was much less tangible but just as debilitating. If someone using the compiler failed to submit a card with the statement END at the back of a program deck (a fairly common error, particularly among student programmers), the program malfunctioned. Ac-

1

cording to one of the evaluators: "The [computer] system went wild. It erased core memory; it erased the master tape; it erased everything." Thus, the occurrence of a relatively minor and common mistake in using the computer effectively incapacitated it until a difficult and time-consuming recovery process could be performed.

The graduate students assigned to the compiler evaluation project carefully documented the bug and sent the information to the vendor. One of them reported: "It was easy; we just had the computer print out a core dump (showing what was stored in the computer's main memory): It was all zeroes." When the vendor eventually released the software for sale, the students eagerly checked to see how the vendor had fixed the error. They ran a program without an END card and discovered, to their dismay, that the program operated exactly as it had before.

The evaluators were furious about their wasted efforts until one of them examined the documentation supplied with the compiler program. Near the back of the manual was a small note in a section titled "Special Features," which allegedly read something like this: "If by chance you ever want to erase core memory and the master tape, merely submit a program without an END card." This announcement was reportedly followed by two footnotes, perhaps apocryphal:

> Note 1   Empirical evidence has shown that it is more efficient to use a shorter program, rather than a longer one.
>
> Note 2   It is our usual policy to charge for utility programs [i.e., programs that do things like erase tapes]. However, this feature is provided free of charge.

Feature is a common English word, meaning "distinguishing characteristic." In the computer world, however, the word is used so frequently in a specialized way that it seems to have taken on the nature of technical jargon. In every description of computing equipment or software, in every comparison of two or more technologies, the word is used, referring not simply to a neutral distinguishing characteristic but to a beneficial one, an advantage.

Rather than fix the problem, then, the vendor had tried to disguise is as a benefit to those who would use it. Unfortunately, however, the problems that this particular bug were likely to create would far outweigh the trivial benefits of the feature.

The students drew a number of conclusions from this experience, some relating to the perversity of vendors, others to the facts of life of building complex programs: It is easier to change the documentation of the program than to change the program. One inescapable conclusion is responsible for the survival of the story as a cultural theme at MIT: What looks like a problem can be transformed into a solution with a judicious change of perspective. Unfortunately, the converse is also true for computer systems: What looks like a solution may often turn out to be a problem. The purpose of this book is to identify the bugs that arise when systems, computerized or otherwise, are designed and used in organizational settings, to describe and explain their causes and to help those who use and design them create systems that improve performance.

RESISTANCE AND THE PEOPLE PROBLEM

When designers look at systems, their perceptions are colored by the pride of authorship. While they may be painfully aware of the technological inadequacies of their creations, they are usually justified in their satisfaction with the results achieved in spite of the time and budget restrictions placed on them. This frame of mind heightens awareness of incidents in which the system is not used as the designer envisioned or intended. It is not surprising that designers often come to view the ways in which systems are or are not used as the major problem related to systems.

The Problem

What catches the designer's eye are situations like the following: After a three-day training course in the use of a word-processing system, secretaries went back to their desks and their typewriters. The designer sees that the word-processing system is not being used but does not see that, in order to use it, secretaries must leave their desks and phones, walk to a room down the hall and share equipment with several other people. Or, two years after the successful conversion to a new financial reporting system, accountants were still using the old system as backup. The designer sees a failure to trust the new system but does not see its periodic downtime during month-end closing. Or, the head of the scheduling department steadfastly refused to support development of a computerized scheduling system that is well within the state of the art. The designer sees irrational fear of change but does not see the very rational career concerns of someone who has achieved success, esteem and a large salary on the basis of mental gymnastics.

   Active misuse, or abuse, of systems can be added to these incidents of nonuse. People have been known to sabotage software by implanting erase or self-destruct instructions. People have used systems to embezzle funds or to make their year-end performance numbers look better. People have apparently used a system as they were supposed to while also finding ways to beat it and to accomplish very different objectives. Numerous incidents of nonuse and abuse are described throughout the book.

The Explanation

Systems can be used (or not used) in so many ways other than those intended by the designers that such uses have been collectively labeled *resistance*. Unfortunately, the term is applied with little discrimination. It has been used in cases where people clearly exhibited the intent to resist—e.g., by physical assault of computing equipment or sabotage of systems. It has also been applied in many cases where the behavior could just as easily reflect ignorance or apathy as active hostility or opposition. This indiscriminate labeling of unintended system use as resistance is a natural outgrowth of the designer's world view. However, it may lead to solving symptoms of the problem rather than the problems and to the development of ineffectual solutions.

There is no doubt that unused systems represent a considerable waste of organizational resources. Time and money have been expended on their development, and management must be concerned about waste. However, some systems create such great costs when they are used that it may be better to lose the initial investment than to insist on using them. The case of Financial Information System (FIS), discussed in chapter 4, is such an example. This system was resisted by a subset of its users for reasons that are quite sensible if viewed in political terms. In spite of several attempts to fix the system, these users remained dissatisfied with it and, over the course of a four-year period, wrote memos about it, discussed it at length informally and attended countless hours of meetings to decide what to do about it.

These users were middle- and high-level accountants and financial managers. The cost to the organization of their salaries during these resistance activities probably exceeded the cost of developing the system by a large margin. Given the costs, it is by no means a foregone conclusion that their resistance was the problem that had to be solved. A different viewpoint sees the problem as the way this system strained the organization, generating the resistance. In this case, trying to overcome the resistance would be solving the wrong problem.

In other cases, systems may provide benefits to an organization even if they are not used. For example, scheduling a fleet of vehicles is a complex and difficult process. Mathematical scheduling models have been developed, and some of these have been computerized. However, if the fleet is large and there are many constraints on the solution, a computerized scheduling system may consume too much of the computing resource for routine use. Nevertheless, if scheduling has always been done manually, management may be unsure that anything close to an optimal schedule has been found. In this situation, developing the computer model may teach the organization how to improve its manual processes. Surely this result would be worth the investment even if the system was never used again. Attempting to see that this scheduling model was used would be misplaced effort.

The Solution

Besides directing attention away from deeper issues, a focus on resistance as the problem may lead to the development of ineffectual solutions. The very term *resistance* suggests that the problem is one of people. Some quirk of personality or some irrationality must be responsible for making some people oppose technology when others accept it so avidly. Perhaps it is a question of age and prior experience. One computer professional told me he predicts resistance by "the rule of 65. If you were in college before 1965, you were not exposed to computers and are threatened by them. After 1965, no problem." Another possible explanation concerns one's left- or right-brain orientation. Intuitive right-brain thinkers have no use for systems, and analytic left-brain people find them indispensable. In any case, a focus on resistance finds explanations in people and personalities.

This, in turn, suggests that solutions are to be found in the techniques for inducing change in people. Examples of these techniques include computer literacy training in grade schools, high schools and colleges; system-specific training in organizations; coercive tactics that leave users no choice about the system (e.g., tying salary or wage payments to system use); and bribery (e.g., economic inducements for system use). Encouraging user participation in system design is also a technique for changing the individual through increased knowledge of system capabilities and increased commitment to its success.

Unfortunately, there is little reason to believe that these solutions are sufficient to overcome resistance or are universally effective in preventing it, as later chapters show. Perhaps it is simply irrational to expect that training will transform irrational or resisting people into satisfied system users.

SYSTEMS AND HASSLES

Users worry not about whether systems are used but about whether they are useful. This frame of mind heightens their awareness of system operating quirks or needless steps that appear to have been designed expressly to save work for the computer or the programmer. It is not surprising that users tend to focus on the hassles of system use as the major system-related problems.

The Problem

Certain situations loom large in the user's view. For example, having developed a system tailored to its unique needs, a department was requested to adopt a slightly different one that made it compatible with other departments. The new system did not supply the reports the department routinely used. Another situation is a company that developed a new cost-accounting system that ran on different equipment than the feeder systems like accounts payable. The accountants were directed to transfer the data manually from one computerized system to another. In another case, computer professionals delivered long-awaited equipment and software in June, but operating manuals and training programs were not available until two months later. Anyone who has lived through one or more episodes like these can perhaps be excused for the belief that systems are more trouble than they are worth.

The Explanation

As sensitive as they are to these hassles of system use, users do not see the conditions that give rise to them. Like many designers, they may view the hassles as exceptional events, problems that will vanish as soon as more time and money are directed toward them (e.g., as soon as a large computer is installed, as soon as a data base management system is purchased, as soon as more programmers are hired). Unfortunately, an understanding of the organizational processes for producing and maintaining systems suggests that the hassles never really go away. They recur, albeit in different forms, and are an inherent feature (or bug) of systems (Kling and Scacchi, 1979).

In spite of this, the quest for the hassle-free system goes on unabated. The grail has been renamed the user-friendly system. An exclusive focus on hassles or user friendliness, however, like an overemphasis on resistance, can disguise bigger problems and can foster inappropriate solutions.

For example, one way to keep the hassles of using a system to a minimum is to localize them. Rather than having everyone in a department irritated once a day, many organizations design the specialized job of system operator. The majority of employees are insulated from system-related problems, but the cost is sometimes the loss of a task that made their jobs complete, understandable or meaningful.

In other cases, the attempt to make systems simple creates simplistic systems. Automated performance monitoring and evaluation systems are not yet capable of qualitative, subjective discriminations, but their output is simple to comprehend. Many organizations take advantage of this opportunity to restructure performance evaluation around criteria that systems are capable of measuring. Here, it seems, the organizations and the people evaluated might be better off with a more flexible, if difficult to use, system.

The Solution

Besides directing attention away from issues like the design of jobs and the standards systems convey to people, a focus on system hassles may create solutions that do not work. The notion of user friendliness frames problems and their solutions in terms of the features of systems and of the delivery mechanisms for them. So, for example, if users have problems, it must be a question of computer hardware: Poor equipment reliability causes frequent downtime; inefficiency causes poor response time. In contrast, it could be a question of software: Users prefer interactive processing to batch. Again, it could be the people and procedures who deliver systems and services: Was a user liaison, a chauffeur or an information center available to help the user? Of course, human engineering factors could also be involved: Perhaps a light pen would have been a better interface device than a mouse. In any case, the presence or absence of some feature of system design is identified as the cause of system hassles.

The implications of a focus on user friendliness for the actions required to prevent or remedy hassles are clear: Design better systems or change the design features of bad systems. This requires well-trained designers who understand the usage implications of various design features and who can make the relevant cost/benefit trade-offs. Thus, the appropriate solutions include training for designers and continuous upgrading and enhancing of systems as hassles are experienced or as newer technology becomes available.

The solutions derived in this way are clear and unambiguous. They specify what actions to take each time because of the underlying assumption that certain design features invariably produce certain hassles. For example, this perspective argues that a two-minute response time is always unacceptable for a terminal user, and the lack of clear performance standards will always create confusion among the people whose performance is measured by a system. In its most extreme form, this perspective assumes that the hassles are not only invariant but also inevitable. Argyris (1971), for example, has argued that certain consequences of systems like feelings of powerlessness experienced by the users occur not only when the systems are badly designed but even when they accomplish the technical and organizational objectives the designers had for them.

The hitch is that the assumptions of invariance and inevitability simply do not fit the facts. For every system in which a design feature produces hassles for users, there is another in which the same design feature exists without evident problems. Sometimes the very same system that creates hassles in one setting will achieve hassle-free benefits in another. As chapter 3 shows, design features alone do not determine the impacts of systems on people or organizations.

IMPACTS AND INTERACTION

Designers tend to see the problem with a system as user resistance; users see the problem as the hassles of system use. A third way to look at system-related problems takes the viewpoint of neither the designer nor the user but enables an understanding of both. This third perspective looks at systems against the background of the total organizational setting in which they are embedded. Peopling this landscape are users and designers, each with their own interests and concerns.

The problems that come into focus against this background are the changes wrought by systems in organizational life. Some of these changes or impacts are related to resistance, which is a condition of the relationship between the users and those who are recommending the system. Other impacts are related to hassles, which are often a fallout of the procedures and structures of the system resource within the larger organization. Yet other impacts are related to organizational features that rarely enter into the awareness of thwarted designers or hassled users.

A focus on system impacts is important for two reasons. First, the impact of a system is, however loosely, related to its objective or purpose. Systems are not built for the sake of having them used or to keep people busy running them; they are intended to achieve some goal, to improve performance. When the desired goal is not achieved or is achieved at a greater than anticipated cost, managers become concerned. Second, systems have been hypothesized, and occasionally demonstrated, to have a wide variety of impacts. The impact may affect the individual, as in job loss; the organization, as in a change in organizational structure; or the relationship between the organization and its external environment, as in gaining a competitive edge. Impacts may occur on the dimensions of job design, job satisfaction, organizational culture, location of work, time frame or communication patterns.

Further, the impacts may be perceived as positive, negative or neutral, depending upon the perspectives of the parties evaluating them.

The Problem

What can be seen with a focus on system impacts are situations like the following: A newspaper systematically pursued the automation of typesetting and printing jobs over a fifteen-year period. At the end of that time, it became apparent that the once predominantly blue-collar work force of the newspaper was now predominantly white collar. Had its industrial relations policies adjusted to this impact? Another example is a conglomerate of entrepreneurial companies that installed a performance-measuring system in all its companies. The measures reinforced conservatism and a short-term profit orientation, and several of the companies lost their leads in rapidly growing markets. Finally, a military organization installed an advanced communication system, causing junior officers to bypass the traditional chain of command.

Such changes transcend hassles and unintended use. Resistance and hassles are both valid ways of viewing problems with systems, but each represents only part of the story. A focus on organizational changes takes both views into account because whether or how the users use systems and the specific features of the systems they use are both relevant to understanding the impact systems have. Consequently, the perspective presented in the book views the problems with systems in terms of their impacts on the people who use them and the organizations in which they are used.

The Explanation

The explanation for the impacts of systems can be found not in the characteristics of users or in the features of systems but in the interaction of system and organizational features. One organizational feature certainly is the characteristics of system users, but many others exist. Another is the characteristics of designers. Still another is the relationships among users and designers.

At least four relevant features of these relationships occur among the users, among the designers and between the users and designers of systems: (1) technology, (2) structure, (3) culture and (4) politics. *Technology* refers to the collection of methods, techniques and know-how for performing a specific task. Users have technologies for accounting, for metal forming, for car making and for recruiting, among many others. Designers have technologies for system building, operations and maintenance. *Structure* refers to the formal patterns of authority and responsibility that prescribe roles and divide the labor among various parties. *Culture* refers to the norms, values, beliefs and nonprescribed behavior patterns that characterize relationships inside and between groups or departments in organizations. *Politics* refers to the processes of negotiation that occur among individuals and groups with differing interests and objectives and with differing bases and relative levels of power. These four features combine with the characteristics of different groups of users and designers to form the context of system design and use.

Systems consist of precisely the same types of features as their context. Systems are nothing more than procedures or technologies that have been made explicit or formalized in some way. In the process of formalization, a structure is created, allocating roles and dividing labor. As the system is used, a pattern of behavior and negotiation is set into motion, enacting a culture and political relationships.

The explanation of system impacts in terms of the interaction between organizational and system features may be called the interaction perspective. This interaction perspective addresses problems that arise at every point in the life of the system. It explains the intentions of system designers (often not computer or systems professionals) in terms of their historical patterns of interaction with users. How well a completed system reflects the intentions of the designers is explained in terms of the interaction between those who commissioned it and the professionals who built it. Resistance to, and the impacts of, a system result from the interactions between design features and organizational context. These phases of system life—intention, execution, adoption and use—are interrelated through the ongoing dynamics of the organization.

The Solution

In its implications for action, the interaction perspective is fundamentally different than explanations based on resistance or on hassles, used alone or in combination. Taken together, the two approaches suggest simultaneous, but independent, efforts to improve system designs while encouraging users to use systems. This assumes that users' behaviors are essentially unrelated to the specific design features of systems. In contrast, the interaction perspective argues that users' behaviors and system design features are tightly linked and that effective system design and implementation strategies must take this into account.

The features—technology, structure, culture and politics—of a system may match those of the larger organizational context, or they may conflict with them. For example, a U.S. manufacturing firm may rightly believe that the kanban system, whereby minimal inventories of raw materials are maintained and ordering closely follows daily production plans, contributes substantially to the success of its Japanese competitors. In attempting to borrow this approach, however, the U.S. firm may find that the kanban method conflicts with its assembly line technology, with its established relationships with suppliers and with the patterned social interactions between workers and managers. Systems are easily transplantable, but they do not necessarily fit the features of the settings to which they are applied.

This is not to say, however, that home-grown systems are necessarily more likely to match their settings than transplanted systems. In many cases, new systems are designed and built precisely in order to change established technologies, structures, cultures and even politics. The interaction perspective suggests that the greater the divergence between a system and its setting, the more likely it is to be rejected or modified, reducing its ability to achieve significant change. The likelihood that a system will have a major impact on its setting is increased to the extent that simultaneous changes are made in other organizational features. In short, the interaction perspective assumes that the impacts of systems are organizational changes and that planning for organizational change requires an approach quite different than the usual methods of system analysis, design and implementation.

CONCLUSION

The bugs and features story at the beginning of this chapter showed, among other things, that different parties can view the same situation in very different ways. What users perceive as a bug, designers and vendors may perceive (or describe) as a feature, and vice versa. Further, different parties identify and are concerned about different kinds of problems with systems. For example, designers focus on the problem of system failure, a major source of which is the failure of users to use the system as the designers intended. Users, in contrast, focus on the problem of systems that don't do what they are supposed to. Another view of problems with systems singles out their impacts on organizational life, regardless of how they are used and how well they were designed.

Perspectives are important not only because they identify the problems that need to be solved but also because they embody a theory of what causes the problems and what needs to be done to prevent them or to correct them. By shifting the viewpoint from that of the user or the designer to one that sees impacts against a backdrop of organizational setting, explanations of problems and prescribed solutions also shift. The problem of resistance must be explained by reference to users' personalities and attitudes and must be solved by training, persuading, coercing or co-opting them. The problem of system hassles is caused by system design features and delivery mechanisms and can only be solved by improving these directly or by training designers to do so. However, if the important problems are impacts, not resistance or hassles, then the explanation is to be found in interactions between system and setting, and the implications for change include a multitude of interventions for bringing system and setting into alignment with each other. The task of this book is to convince readers of the validity and utility of the interaction perspective on the problems of systems.

2

System Design Features

INTRODUCTION

The thesis of this book is that the impacts of systems are produced by the interaction of system design features with features of the organizations in which systems are used. Therefore, the ability to identify the relevant system design features and the interacting organizational features is essential to explaining, predicting or controlling system impacts.

The diversity of systems makes it desirable to classify systems into types with different design features. Many commonly used classifications of systems are based on dimensions that are not strongly related to organizational impacts. For example, systems are often classified by the type of technology they employ such as nonautomated, automated with batch-processing technology or automated with interactive-processing technology. However, these types offer few clues to the likely impacts of systems on dimensions such as organizational structure or job satisfaction. Another common basis of classification is system users such as clerical personnel, engineers and other professionals or technicians, middle managers and chief executives. But systems are often used by several user groups and may affect each group in different ways.

The classification scheme used here is based upon the function a system serves in the organization in which it is used. Five types of system functions are identified:

1. To structure work,
2. To evaluate performance and motivate people,
3. To support intellectual processes,
4. To augment human communication,
5. To facilitate interorganizational transactions.

Figure 2.1 presents each type of function and a descriptive title for it. The titles assigned the first three system types are identical or similar to those found in most typologies of management information systems (e.g., Gorry and Scott Morton, 1971). It is important to recognize, however, that other typologies have different bases of classification and define their types in different ways.

Figure 2.2 presents the design features that characterize system types and that help identify their organizational impacts. Several points should be made about the design features listed in figure 2.2. First, the features are independent of technology. An operational system may be unautomated or automated using any type of technology. Second, the features do not assume a particular type of user. Some operational systems may be used by chief executive officers, and some planning and decision systems may be used by technicians and clerks. Third, some systems are composed of subsystems of multiple types. For example, a manufacturing system may have operational, monitoring and control and planning and decision subsystems. While such a system cannot be neatly classified into a single type, the collection of design features it encompasses is a useful guide to its likely organizational impacts.

FIGURE 2.1
System Types and Functions

System Types	System Functions
Operational	To structure work
Monitoring and control	To evaluate performance and motivate people
Planning and decision	To support intellectual processes
Communication	To augment human communication
Interorganizational	To facilitate interorganizational transactions

FIGURE 2.2
System Types and Design Features

System Types	Key Design Features
Operational	Work rationalization
	Work routinization
Monitoring and control	Standards
	Measures
	Evaluation
	Feedback
	Reward
Planning and decision	Models
	Data analysis and presentation
Communication	Communication procedures
	Communication mediation
Interorganizational	Structuring or mediation of interorganizational transactions

The remainder of this chapter explores each system type in turn. The key design features are described and related to the corresponding features of the system's setting. Examples are given to illustrate the design features and to demonstrate how the typology can be used to classify other systems.

OPERATIONAL SYSTEMS

Operational systems serve the purpose of structuring the performance of work tasks. Typically, operational systems focus on the physical or tangible aspects of an activity, such as material handling, the preparation of drawings or designs and document production. However, the intellectual aspects of the activity may also be affected, usually by altering the range of choices that need to be made or the alternatives that need to be considered. When an activity is primarily intellectual, with a very limited physical component, the type of system applied to it usually falls into the planning and decision category. Examples of primarily intellectual activities include medical diagnosis, strategic planning and investment evaluations.

The task structured by an operational system may be performed by people in any occupational category or hierarchical level. Operational systems structure the work of clerks and secretaries, blue-collar workers, engineers and architects, accountants and managers. The tasks structured by operational systems may be central or peripheral to the core business of an organization. Operational systems may structure the processing of claims in insurance companies or the billing of advertising customers in a newspaper.

The key design features of operational systems are work rationalization and work routinization. *Rationalization* means reorganization with up-to-date methods and procedures. This reorganization is usually intended to make the work process more efficient—that is, to achieve a better ratio of outputs (or products) to inputs, which may include labor, machines and materials. In many work processes, labor is a major component of production cost, and the goal of rationalization is to reduce the labor time required to make a product and, hence, to reduce cost.

Routinization means to make the work process uniform so that outputs are consistent and predictable quantities and qualities of inputs—labor and materials—can be used. An efficient work process has not only a good yield of outputs to inputs but also acceptable results most of the time. In a production environment, as opposed to a research lab, occasional excellent results are not well regarded if 90 percent of the product is of substandard quality. An efficient work process also produces outputs of predictably acceptable quality regardless of the person or machine performing the work. Small allowances may be made for starting up a new machine or for training a new employee, but the goal is usually to obtain comparable results from every unit of production.

The design features of operational systems are closely related to certain features of the organizational setting in which the work process occurs. Because operational systems are intended to reduce labor costs, a relevant feature is the size, composition and structure of the labor force for that production process. In addition, because operational systems impose new methods and routines on the performance of work, they affect the two basic building blocks of organizational structure—the design of jobs and the coordination of work flow among jobs. The latter, in turn, is associated with culture, the patterned interrelationships among people and organizational units.

Two examples of operational systems follow, a letter of credit system in a bank and computer-integrated manufacturing. These examples illustrate the rationalization and routinization of work and describe enough of their organizational contexts to demonstrate their relationship to the features of work force structure, job design and work flow coordination and organizational culture.

Letter of Credit System Example

Mattheis (1979) offers a graphic description of an operational system in a bank, showing the relationship between systems and the features of work force composition, job design and work flow coordination. The system in question is letter of credit processing. Letters of credit are guarantees of payment by a bank to a third party from whom the customer is purchasing goods. These financial instruments are widely used to finance shipments of goods in international trade. The complexity of the terms and conditions entailed in a letter of credit make processing difficult: The customer can be expected to amend the letter at least once before payment takes place.

Prior to 1975, Citibank did not differentiate letter of credit processing from the dozens of other transactions it performed in its operations headquarters. Requests for letters of credit went into a processing chain along with checks, deposits, stocks, bonds, loans, money transfers and collections. What happened then to the requested letter of credit is described in the following excerpt:

> Sorting clerks in the mailroom check the contents of the envelope and send the item on to the bank's letter of credit department on the building's 24th floor. The item lands on the desk of a preprocessing clerk who determines the source of the item, whether it is from a Citibank branch, a correspondent bank, or a government agency. The clerk shunts the item to the correspondent bank section. There, another clerk has to determine whether the item requests that a letter of credit be issued, amended, or paid, or whether it is a customer inquiry. The clerk sends the Lyons item to the issuance unit.
>
> Three days later, the credit for the Lyons company's shipment is issued. At least 14 people—a typist, a log-in clerk, a preparer, a signature control clerk, two checkers, the department manager, a central liabilities clerk at the bank's uptown corporate headquarters, a marketing officer who approves the credit (also from uptown), the accounting department, the files unit, and the customer service clerk—have acted on it.
>
> The original source document from the Lyons correspondent has been read, reread, checked, and rechecked. It has been crumpled, clipped, stapled, unstapled, rubber-banded in bundles of cards and tickets, stuffed into envelopes, copied, copied from, annotated, and preserved in a cardboard file folder.
>
> The processing of the credit issuance has generated a stack of papers maintained in some six files. An offering ticket has gone to central liabilities and to the marketing officer, with a copy for the file; a five-copy fanfold has been typed, split, and its folds dispersed to a variety of destinations;

accounting tapes, proof tapes, and MIS tapes have been punched, rolled, and delivered around the bank; special instructions have been duly noted and recorded in duplicate, one set concerning the Lyons customer, and the other the beneficiary out in Michigan.

The Diebold files are bulging, and the department requisitions additional forms and paper clips. Yet a week after the credit is issued, the customer requests that it be amended, starting the same process all over again. [Mattheis, 1979: 146–147]

Ultimately, Citibank replaced this operational system with another:

It is now October 1976, a year and a half later; the same company requests a similar credit through the same ... correspondent. In the Citibank mailroom, a clerk routes the item to: Letter of Credit, 21st floor, European Division. At the division's preprocessing desk, a clerk notes the item's source and puts it in a slot marked "France" on an automatic delivery cart, a mail "robot." As the robot moves along the aisle, the item is plucked from its slot by an individual sitting in a neat, cockpitlike work station, where she is typing onto a CRT terminal keyboard. As she punches the last key, a printer terminal nearby prints out a formal letter of credit. It is mailed that day. She immediately puts the customer's original request and instructions on microfiche and stores it in a small file case on the flat top of the work station at her right hand.

Where it once took days, 30-odd separate processing steps, 14 people, and a variety of forms, tickets, and file folders to process a single letter of credit, it now requires one individual less than a day to receive, issue, and mail out a letter of credit—all via a terminal that is fully online to a minicomputer-based system. [Mattheis, 1979: 147]

Mattheis credits the new system with having achieved many benefits for the bank, among them improving customer service, decreasing the amount of time (and presumably the labor cost) required to process a letter and increasing the job satisfaction for the clerks in this department.

Computer-Integrated Manufacturing Example

Computer-integrated manufacturing represents the linking of operational systems for the design process with operational systems for manufacturing. The former are sometimes called CAD for computer-aided design, and the latter are called CAM for computer-assisted manufacturing.

CAD does not automate all the knowledge and skills of those engineers who design such things as industrial machinery or computer logic and memory circuits. However, CAD is capable of automating parts of the design process—for example, drafting layouts and performing engineering analysis. In addition, CAD may structure the design process and guide the engineer through it, enabling individuals with fewer skills or lower skills to produce results comparable to or better than those of highly trained and experienced professionals. Consider this description:

> "Everything a draftsman needs is in the computer," says Ronald A. Cenowa, a computer engineer in GM's Fisher Body division in Warren, Michigan. "Anything that a draftsman conventionally does using triangles, pencils, compasses, and so on will be done mathematically within this system." CAD not only speeds up the slow and laborious work of drafting, but also enables the designer to study various aspects of an object or assemblage by rotating it on the computer screen, separating it into segments, or enlarging or shrinking details. . . . What's more, if the necessary programming has been done, the designer can analyze and test the things he designs right in front of his eyes, subjecting them to electronically simulated temperature changes, mechanical stresses, and other conditions that might impinge in real life. [Bylinsky, 1981: 106]

CAM refers to the computer control of processes, such as chemical refining or steel making, or of machines, such as robots and machine tools. In CAM, a computer monitors a production process or a machine for variances from designated standards. Then the computer either triggers the physical devices that will correct these deviations or communicates the deviations to a human operator for corrective action. CAM, on the one hand, can be thought of as occupying the opposite extreme of a continuum that has fixed automation at one end. Fixed automation refers to machinery that is capable of producing only one kind of output—e.g., drill presses or the equipment that molds automobile bodies. CAM, on the other hand, is a flexible technology capable both of producing different responses, depending on production variances, and of being applied to different production processes. An example of this flexibility is an industrial robot that can be programmed to drill holes today and reprogrammed to spray paint tomorrow. The numerical control of machine tools probably represents a midpoint on the fixed automation-CAM continuum. In numerical control, a set of software instructions is encoded on paper or magnetic tape to control the operation of machine tools. Numerical control provides computerized control of production variances without providing the flexibility of robotics.

CAM, at least in the form of numerical control, has become relatively commonplace in industry, but the integration of CAM with CAD is less widespread. This integration is expected to produce payoffs so significant that many sources have proclaimed the onset of a new industrial revolution. For example:

> Pratt and Whitney Aircraft, one of the companies that have gone furthest in applying CAD/CAM to manufacturing, now makes turbine blades, among other things, directly from CAD drawings, with the whole process automated. Edwin N. Nison, manager of technical and management data systems and scientific analysis, says that in many cases "thanks to CAD we have gained a 5-to-1 or 6-to-1 reduction in labor and at least a 2-to-1 reduction in lead time. And these ratios go up as high as 30 to 1 and 50 to 1 where we linked CAD with CAM." [Bylinsky, 1981: 107–108]

This reduction in lead time is derived, at least in part, from a reduction in labor hours. Achieving these results requires a process of rationalization and routinization focused on the organizational features of job design and work flow coordination.

> The achievement of something like full ... [CAD/CAM] entails thoroughgoing analysis of manufacturing operations—and perhaps restructuring of them—in order to translate them into precise computer language. This is a demanding and time-consuming task. ... Factory operations may seem orderly enough until you try to describe them in computer programs, but then they begin to look quite irregular. ... A major [obstacle] is the lack of uniformity in procedures from one plant to another, or even from one technician to another in a given plant. [Bylinsky, 1981: 109]

MONITORING AND CONTROL SYSTEMS

Monitoring and control systems are intended to evaluate the performance of people and/or organizational units and to motivate people to improve their performance. The dimensions of performance monitored and evaluated by these systems may be objective, like dollars spent, or subjective, like package appearance. They may relate to the outcome, like the number of units produced, or to the behaviors required to produce outcomes, such as patient contact hours or billable hours (Ouchi and Johnson, 1978).

The performance in question may be that of an individual from any hierarchical level or occupational group or it may be that of a department, division or subsidiary. The purpose of the monitoring may be to assist in the future planning process, to facilitate learning about controllable variances, to maintain performance within certain parameters or to motivate people to achieve better results in the future (Handy, 1976). The classic examples of control systems include budgeting, responsibility center accounting and standard cost accounting. Finally, the focus of control may be internal or external. Thus, the system may be designed to assist the person or unit in self-control—that is, setting personal goals and monitoring achievement—or for external control—that is, administering rewards or punishments for achieving goals set, at least in part, by others. An example of a self-control system is a diet; an example of an external control system is an audit by the Internal Revenue Service.

According to accounting theorists (e.g., Flamholtz and Tsui, 1980), the key design features of monitoring and control systems include the following:

· Goals or standards on each dimension of performance for which an individual or unit is held responsible: e.g., past performance.
· Measures of actual performance.
· Evaluation: comparison of performance with the goal or standard.
· Feedback: communication of performance against the standards within a time frame to influence subsequent performance.
· Reward: administration of extrinsic rewards like bonuses, commissions, piecework compensation, or merit raises and punishments like pay docking, suspension or termination.

Depending on the purpose of the system and its internal or external focus, some of these elements may be missing, but some form of measurement and communication of results is essential to any monitoring and control system. The design features of a monitoring and control system are derived from the cybernetic model, which may be labeled the thermostat model. Thermostats operate by measuring the temperature, comparing this to a preset standard (68 degrees F in a conservationally minded environment) and then initiating action to achieve a reduction in the variance between actual and standard. The action may be to turn on the heat (if it is a heating system and if the temperature is too cold), to turn on the air conditioning (if it is an air-conditioning system and if the temperature is too hot) or neither (if the temperature is warm enough with a heating system or cold enough with an air-conditioning system). Control theorists have argued that people's behavior can be measured and compared to preset objectives to determine the extent of the undesirable variance, if any. When presented with feedback—that is, evidence of the size of the negative variance—the rational employee (presumably armed with the knowledge of what actions will produce what outcomes) will adjust his or her behavior in ways that bring results back to standard.

These design features of control systems are related to features of the system's organizational context. One of these is the design of jobs—the degree of autonomy, the closeness or slackness of supervision, the stress of high performance demands or the boredom of easily achieved goals. Another feature concerns culture, the ongoing behavior patterns, values and belief systems affected by the system in question.

Two examples, illustrating monitoring and control systems, are presented in the following sections. The first concerns office facilities management in a government agency; the second involves the productivity of clerical employees. The design features of control systems can be identified in these examples, as can their relationships to the features of job design and culture.

Space Management Example

The space management example concerns a control system developed to improve the management of office space in a large U.S. government agency, here called FBT.* FBT employs over 70,000 employees and maintains more than 1,000 local offices across the nation. The agency has a national head office and fifty-eight districts.

FBT's requirements for office and computer operations space are enormous, and the cost of this space is high. Rented facilities range in size from 350 square feet for a local office to 500,000 square feet for a computer-processing center. On an annual basis, the space cost is almost $150 million. Prior to 1972, however, FBT did not manage this cost. Each government agency applied to the General Services Administration (GSA), the procurement arm of the federal government, when it needed space; GSA assigned space upon request and had the responsibility for controlling costs.

Obviously, this procedure gave agencies no incentive to economize. In 1972, the Public Buildings Act attempted to shift the space management burden back to the agencies by requiring them to reimburse GSA for space used and to defend their own space budgets to Congress. The Public Buildings Act effectively made GSA a landlord to FBT and other agencies. These changes radically increased the importance of space management within FBT, because after 1972, the space budget became the second-largest budget item (salaries were the largest).

GSA maintained a system with data about the amount and cost of space used by each agency. With this system, GSA computed a standard level user charge (SLUC) that an agency was required to reimburse. FBT managers doubted the accuracy of these charges and believed they did not have enough information about their space utilization to achieve efficient control over it. In this culture of scrutiny and suspicion, FBT needed a way to hold its own. Accordingly, FBT developed a space management system for its own use.

---

*The data for this case were collected by George Conley for a course assignment at MIT's Sloan School of Management in spring 1981.

Employees from the local, regional and national levels of the agency all participated readily in the development of a tool that would benefit them in their dealings with GSA. At the end of the two-year development process, in 1975, employees in the regions actively cooperated with FBT headquarters personnel in creating a data base containing current data about buildings occupied, the nature of space held within the building, the number and function of employees occupying the space and historical data on these dimensions for the preceding five years.

The system, known as SMIRS (Space Management Information Retrieval System), computed several measures of space management performance. One set of indicators concerned cost: total cost per square foot, total annual costs, cost to date and cost per employee. Another was the space utilization rate, figured as the total square footage of office space divided by the number of employees occupying it. The third performance measure was an ordinal ranking: The system was programmed to produce listings of agency districts in order of performance on any dimension such as total square footage, total cost, number of employees, utilization rate and date of lease expiration.

In keeping with the original purpose of the system—to insure the accuracy of space charges paid to GSA—SMIRS produced quarterly reports of the discrepancies between its computations and SLUC billings. As a space management tool, SMIRS produced a national guide of office locations, a report of annual lease expirations and a quarterly office profile with complete information on space, personnel and cost for each location, totaled by district, region and nationally. In addition, the system produced a utilization rates report that ranked the fifty-eight districts in ascending order of utilization rate (space per employee).

Typing Productivity Example

A simple example of a control system is the feature on many word-processing systems to count the number of keystrokes per unit of time. One can envision a novice data entry person using this feature for self-control purposes to measure learning and to set goals for improvement each day. In fact, however, systems like this rarely report statistics to the person measured. Instead, the data are collected in a management report for inspection by a supervisor who uses a specially programmed terminal. These data frequently form part of the job evaluation and incentive payment schemes for typists and data entry clerks. When used in this way, the keystroke counting systems can dramatically increase the pressure experienced by the workers, leading to stress, reduced job satisfaction, high turnover and other problems. To avoid some of these problems, some organizations have publicly disabled the counters.

It is interesting to observe that the same issue arose with manual typewriters in the early years of this century, when industrial engineers first turned their talents toward clerical work, as illustrated by the following excerpt:

> Stenographic output and other forms of typing were studied most carefully. 'Some typewriter concerns equip their machines with a mechanical contrivance which automatically counts the strokes made on the typewriter and records them on a dial.' This meter was used in conjunction with a time clock, which the typist punched at the start and finish of each job. Metering of this kind was used as the basis for piecework payments (it took some time before management experts discovered that under such a regimen typists never used the tabulator key, always the space bar, in order to increase their count). [Braverman, 1974: 307–308]

PLANNING AND DECISION SYSTEMS

Planning and decision systems are designed to support processes and activities that are primarily intellectual, such as drawing conclusions from evidence, making predictions from past performance and deciding on an appropriate course of action to follow. The intellectual problems attacked by these systems may arise so frequently that the benefits of routinization are sought or they may be one-of-a-kind situations of such importance that steps are taken to improve the quality of the decision outcome.

The plans and decisions subjected to systemization may be those habitually made by relatively low-level employees in the organization like the inventory controllers who determine when and what quantity of parts and materials to reorder. Planning and decision systems, however, have also been applied to the responsibilities of high-ranking staff specialists and professionals. Structured decision methods and computer-based programs have been designed to support the thought processes of investment analysts and stock portfolio managers, engineers and geologists, brand managers and marketing experts and physicians, among others. Some planning and decision systems are directed at matters of concern to the highest managers in an organization: for example, the decision to make an acquisition or divestiture, the formulation or modification of marketing strategy, the choice of location for a major new facility and so forth.

The key design features of planning and decision systems are models and data manipulation. In different forms, these features are present in virtually every planning and decision system. Models are formalized descriptions. They describe either a real-world process that an expert is attempting to analyze—for example, the economy or a company's financial performance—or the mental processes that an expert would use in attempting to understand the real-world process. Models of the former type may be optimizing, designed to recommend the best course of action subject to specific constraints, or simulation models, designed to evaluate the effects of alternative actions under differing environmental conditions. The discipline of operations research has provided the base of knowledge that underlies the construction of models of this sort.

Models of the latter type are often called expert systems because they simulate the analysis an expert would make. They lead the user of the system through a chain of questions and answers, requesting the user to supply the data needed at each step of the analysis and indicating which alternatives can be ruled out when the data are analyzed. Knowledge about how to construct models of this type has been provided by a branch of computer science known as artificial intelligence.

In certain situations facing managers and professionals, operations research or artificial intelligence modeling is inappropriate. No operations research model may exist for the problem, the data it requires may be unavailable or the assumptions and constraints on the solutions may be unrealistic. In addition, no expert may be available whose decision processes can be observed and modeled. Finally, it may be unclear whether a problem exists and what its precise nature is. In cases like these, it may be possible to construct decision support systems that help people by manipulating available data and by presenting them in a format that facilitates learning and decision making.

Credit for the term *decision support systems* is attributed to Gorry and Scott Morton (1971), who categorized these systems by the type of decision supported (decisions concerning operations, managerial control or strategic planning) and by the degree of structure (unstructured, semistructured or structured) in the decision process. Keen has defined these systems in ways that distinguish them sharply from operations research models:

> Decision support systems are interactive computer systems developed to aid managers in their problem-solving in tasks that involve judgment and thus cannot be automated. [1979]

> Decision support systems are small-scale, interactive systems designed to provide managers with flexible, responsive tools that act in effect as a staff assistant to whom they can delegate more routine parts of their job. . . . They do not impose solutions and methods but provide access to information, models and reports and help extend the managers' scope of analysis. [1980]

That decision support systems embody a particular technology, like interactive computing, is not the key feature of these systems. What is important is that they enhance the learning process, the activity of thinking about organizational problems.

The key design features of planning and decision systems, models and data manipulation, are related to features of the contexts in which the systems are used. Because these systems are intended to facilitate data analysis, they may reduce the amount of labor involved in data handling. This is related to the size and composition of the information-processing work force, which is almost totally white collar in nature. Because these systems seek to impose structure on decision-making and planning activities, they have implications for the design of the decision makers', jobs and the skill requirements for these jobs.

Furthermore, many decision and planning situations involve the input of several parties. In some cases, staff specialists in headquarters make plans and decisions that affect line or staff personnel in divisions or subsidiaries; in others, divisions and subsidiaries have relative autonomy. Systems that structure decision-making and planning processes may encourage rethinking the structural relationships between line and staff, which are usually described in terms of centralization or decentralization. Structure, in turn, is related to organizational culture, patterned behavior and belief systems that characterize the relationships among parties to decision-making and planning processes.

Three brief examples of planning and decision systems follow: one describing an operations research planning model, another describing a decision support system and a third describing an expert system. These examples illustrate the use of models and data manipulation in contexts in which work force characteristics, job design, and organizational structure and culture are salient features.

Modeling Example

This system, Integration of Marketing and Refining (IMR), uses the operations research technique of linear programming to determine the best pattern of marketing, refining and supply activities within the company. The application began as a research activity within the Canadian associate company of British Petroleum (BP) in 1964 (Stewart, 1971). Prior to this time, BP and the Canadian company had developed substantial expertise in operations research but had not applied this directly to their long-range planning efforts.

Three-year plans were formerly produced annually through a process that entailed close collaboration between functional specialists in marketing, central planning, supply and development and refineries in the head office of BP in London and corresponding specialists in the associated companies. The marketing department developed strategies and intelligence about international conditions that might affect sales. Central planning produced economic forecasts. Supply and development made logistical plans for coordinating the movement of crude oil and refined products and for planning refinery capacity increases. Refineries supplied technical information about the capacity of existing plants.

The data produced by these four types of functional specialists were incorporated into the plan independently. Ramifications of each group's data on the assumptions underlying the other groups' input were not considered in detail. For example, a plan to sell a certain amount of one product might reduce the production capability available for other products, yet the refineries might have estimated their production costs on the assumption that multiple products were to be produced in a certain mix.

> For instance, in conventional planning and budgeting the sales growth is usually fixed first, and then the repercussions are worked out on production, distribution, R. & D. and so on. Even these repercussions are not worked out in detail—one often does not know whether a particular product development will cause a bottleneck somewhere in production. [Stewart, 1971: 137]

The IMR was built to consider the inputs of the various functional areas simultaneously rather than sequentially and to determine automatically a plan to maximize profits subject to certain constraints. The Canadian associate company developed a model to do this by augmenting an existing model of the refining process with data from marketing and central planning. When IMR's three-year plan was compared to that produced by the conventional planning process, it was discovered that the IMR model suggested a way to achieve much higher profits. This led to a decision to expand the model so that it could evaluate up to thirty-six policy alternatives over a six-year period. Subsequently, the functional departments in BP's head office were asked to check the model in detail. When this lengthy and expensive process of validation and testing was accomplished to the satisfaction of senior managers, plans were made to implement the IMR model in the planning processes of other associated companies and to adopt a macro IMR model for planning the joint activity of all associated companies.

Decision Support System Example

An example of a system designed with these assumptions is given by Berger and Edelman (1977). It is a decision support system intended to facilitate analysis and problem solving about human resources management at RCA. The system, named IRIS, replaced and augmented a number of unintegrated computer systems that dealt separately with human resource management issues including payroll, pension plan administration, insurance, labor relations, affirmative action programs and in-house telephone directories. The motivation for IRIS came from an unsuccessful attempt by RCA managers to resolve a question on personnel matters with the data available in earlier computer systems. IRIS was designed to permit managers to make ad hoc inquiries on the human resources data base whenever they choose. Furthermore, these questions had not necessarily been anticipated when the system was designed. These and other features of IRIS differ substantially from the preconceived nature of questions that can be asked of optimizing models.

Some of the input data for IRIS come from operational systems like payroll. RCA periodically conducts a personnel survey to obtain new data directly from the people involved and to validate old data. Much of the information stored and manipulated by IRIS is highly confidential (e.g., salaries), so the system has built-in security features like passwords to protect sensitive data. IRIS has a sample data base feature that allows a user to try out a question on a small, random sample of about 100 people (chosen by the system) before trying it on the entire data base of maybe 10,000. This allows a user to learn inexpensively whether the question will yield the desired answer. In addition, the system can be used to project the impact of an intended salary action (say, a 10 percent wage increase for clerical workers in a given job category) on departmental budgets.

IRIS was designed to be used by managers who are not computer professionals. A compact and easily learned language was developed to allow a user to instruct the system in what is to be done (develop a report, change a report, review libraries of reports). Useful report formats can be saved in a library and then recreated with current data whenever desired. Users of the system can specify what data will be saved and when old data will be purged from the data base.

Expert System Example

One of the earliest expert systems was MYCIN, which assists medical professionals in diagnosing and recommending treatment for bacterial infections (Webster and Miner, 1982). A physician sitting at a computer terminal will be prompted by the program to type in the patient's name, age, sex and race. Then the program will ask the physician a series of questions such as "Are any positive cultures obtained from the patient?" or "Has the patient recently had symptoms of persistent headache or other abnormal neurologic symptoms?" Based on the answers supplied, the computer program eliminates various possibilities and eventually converges on a recommended course of treatment.

Other examples of expert systems include a program to program computers, a program to evaluate geologic sites for potential mineral deposits and a program to configure computer systems to meet customer requirements. Expert systems are so called because they imitate the if/then reasoning process used consciously or unconsciously by people who are recognized experts in their fields. A system developer will observe and question one or more experts about their decision-making strategies before programming an expert system. In addition to making the experience and opinions of experts widely available to people who are knowledgeable but not expert, these systems help train the people who use them. Some expert systems are programmed to print out, in response to the user's typed question Why?, the chain of reasoning that led the program to pose a particular question or propose a particular answer.

COMMUNICATION SYSTEMS

Communication systems are designed to augment communication among people. Augmenting interpersonal communication is a secondary function of the other types of systems discussed in this chapter. For example, operational systems communicate the information required for successful work flow coordination in addition to rationalizing and routinizing work. Monitoring and control systems communicate performance data to various parties in addition to measuring and comparing performance to preset standards. Where planning and decision processes involve more than one person, planning and decision systems help facilitate the communication of assumptions, data and interpretations.

Other systems, however, have the augmentation of communication as their primary function. These systems are characterized by two types of design features: procedures intended to reduce the constraints of geography and communication media that change the form, speed or interactiveness of communication. Examples of communication procedures are interoffice or U.S. mail delivery, message-taking activities, the organizational units designed to coordinate message handling and the discipline of checking regularly for messages. Communications media include paper and forms for memos or reports, telephone and voice storage and synthesis, graphical representations and video recordings and transmissions. Recent technological advances have led to a proliferation of communication systems employing computer-based text messaging, videodisc training programs, voice mail and video conferencing, among other technologies. The design features of communication procedures and media are related to features of organizational life, especially those concerning time and geography. Communication systems attempt to transcend distance and make the successful operation of business independent of the location of people and resources.

> The development of technology has never been aimed solely at saving human labor and reducing physical exertion. It has also been aimed at making geographical space, the space inhabited by mankind, *fungible*. If this condition were achieved, an infinity of problems that have always plagued individuals and society would be resolved. The fungibility of space would mean that every point in that space would for all practical purposes be equivalent to any other point. [Gottmann, 1977: 306, emphasis in original]

Another important feature of organization related to the design features of communication systems is the communication network, or pattern of communication channels. Media differ in the number of people brought into contact in a single communication event: With telephone only two are connected, but television may reach millions. They differ in direction of communication: Mail is one way, telephone is two way, and CB radio is many to many. They differ in requiring the simultaneous presence or participation of communicating parties: Mail does not, but telephone does. These media characteristics shape and constrain the patterns of communication among the connected parties. This central feature of organization is strongly associated with the structural aspects of work flow coordination and degree of centralization and with organizational culture.

Two short examples of message systems are presented to illustrate the design features of procedures and mediation as well as some of the related organizational features.

Message System Examples

CPA, Inc. is one of the so-called Big Eight accounting and consulting firms. It has over 100 offices worldwide. The offices largely operate independently of each other; consequently, interoffice communication is adequately handled through the phone and mail systems. In one of its large local offices, however, CPA, Inc. has devoted a great deal of time and attention to designing procedures and systems for internal communication.

CPA, Inc. is in the business of serving the clients of its practices in audit, tax and management consulting. The professionals who provide these services spend much of their time on the premises of client firms. In fact, junior people on the audit staff spend so little time in the local office that they have no permanent offices, desks or telephones assigned to them. Senior people may be out of their offices 70 to 80 percent of the time. Under these conditions, it is difficult for clients to get into direct contact with their intended party at CPA by telephone.

Naturally, CPA would like to be sure that clients get the best possible response when they telephone the office. This means that the called party should always take the call when he or she is in the office unless he or she is meeting with another client; messages should always be taken for a called party who is unavailable to answer the phone; and phone messages should be delivered to the called party as soon as possible. This is the definition of the required level of service that CPA wishes to provide the clients who telephone its office.

This level of service could be achieved in several ways. For example, each professional could be provided with secretaries to answer telephones. Telephones could be outfitted with answering devices. However, CPA has chosen a different approach. Clients are given a single telephone number for all professionals in the midwest office. The number rings at a central switchboard. (The firm's partners have private lines that do not go through the switchboard, but these are not used for incoming client calls.) Console operators at the switchboard check a record of who is in the office that is kept current to the half-hour. If the called party is in the office, the operator rings his or her extension. If no one answers the extension, the operator pages the party over an intercom that can be heard in all offices, hallways, lounges and restrooms. If the party does not answer the page, the caller is transferred to someone in the message center who will take a message. If the called party is not in the building, the telephone operator immediately transfers the call to the message center without paging or ringing the extension.

The message center writes down the messages on slips of paper. Copies of these slips are filed and stored in case they are needed for future reference. The message center also delivers the messages at the first opportunity. This entails calling or paging the called party or his or her secretary on a prearranged schedule or when the party reenters the building.

Obviously, the success of this system requires an accurate record of who is in the building. Here is how this record is maintained. The office has only one (official) entrance and exit. All visitors must use this entrance, and all members of the professional staff are expected to use it to enter and leave the office (unless they are going to rooms on a different floor). As they enter in the morning, leave in the evening, or enter or leave for lunch and client visits, all members of the professional staff file past a receptionist and declare their presence or destination. Usually, they just smile and say "Good morning" or "Lunch" or "XYZ client" to a seasoned receptionist, but when a new person is on duty, they must state their names as well. The receptionist writes down their names and location on an electronic note pad that is connected to devices in the telephone-answering and message center area. Every half hour, someone updates mimeographed personnel lists with vacation and travel plans and the Electrowriter printout. This communication system enables telephone operators to handle calls and the message center to deliver messages.

SOWHAT Oil Corp. has taken a different approach to internal communication. Top management in the corporate research center created a policy that all internal memoranda would be sent via a computer-based message system rather than intrafacility mail. Prior to this policy, internal memoranda at SOWHAT were prepared as follows. Professionals would dictate memos or write them out longhand. These memos would be typed by secretaries; signed, copied and filed; hand carried to their intended recipient and answered and filed there. The new system short circuited several steps in this chain. Copying and hand carrying were eliminated since memos were displayed on the CRTs in each recipient's office. Only one electronic copy of each memo was filed, even if 600 people received it. Filing was done automatically. Indexing and retrieval facilities made it easy for recipients to reread the memo at will. Finally, because all professionals have convenient access to terminals (and less convenient access to secretaries) many of them began to keyboard their own short memos, eliminating the transcription of longhand or dictation by secretaries.

INTERORGANIZATIONAL SYSTEMS

Interorganizational systems are intended to facilitate transactions involving more than one organizational entity. The other types of systems described in this chapter—operational, monitoring and control, planning and decision and communication—concern transactions that occur within an organization, although they may involve interactions among internal subunits. Interorganizational systems, in contrast, concern habitual interactions among autonomous but interdependent organizations. Examples of these interactions include purchases by hospitals from wholesale distributions, the sale of airline tickets by travel agencies and the use of banking services by business customers.

Some firms have developed systems to facilitate these transactions, and these systems may have subsystems with the design features of operational, monitoring and control, planning and decision and communication systems. These design features are related to specific organizational features inside one or more of the parties to the transaction, either the developer of the system or the firms who use the system to conduct transactions with the system developer. However, interorganizational systems also have design features that relate not just to the features of the transacting parties but also to the relationship between them. These design features are the procedures that structure the transaction and the media and technologies that change the nature of the transaction conducted. The related organizational features are the quality and nature of relationships, first, with customers or suppliers and, second, with competitors.

An example of an interorganizational system is presented next: an order entry system for hospitals developed by a wholesale distributor of hospital supplies. Other examples include the systems used by independent insurance agencies to record policy purchases and to make claims with insurers, systems used by travel agencies to make reservations for airline travel with carriers and to issue tickets and the cash management systems developed by banks for corporate customers to use for managing and transferring funds in bank accounts. Each of these systems can be related to features within the firms that develop or use them, but they can also be related to features of the interorganizational relationships between the firm and its customers and suppliers.

Interorganizational System Example

American Hospital Supply is a manufacturer and wholesale distributor of the products consumed in large quantities by hospitals—e.g., syringes, disposable gloves, bandage materials, intravenous solutions. American Hospital Supply has developed an order entry system that makes use of special computer terminals located in the hospital. Hospital customers pay a fee for this service that allows them to send orders quickly and accurately, avoiding busy telephones, slow mails and inaccurate human order entry clerks.

> Hospitals are in trouble. . . .
>
> And that is where the big hospital supply house comes in. [American Hospital Supply] currently is the only company that is offering both automated order-entry and inventory control systems to hospitals. Most of our competitors can provide one or two pieces of the [materials management] puzzle, but only American can provide them all, says Robert M. Waller, director of corporate distribution. The price tag for these services is not cheap—about $1,400 a month for most of the company's customers. American has signed up 3,000 of the nation's 7,000 hospitals for its order-entry system, and by aiming at institutions with at least 300 beds, it has already signed up 43 hospitals for its newer inventory management package.
>
> For order-entry, these hospitals can use their computer terminals to communicate directly via an on-line hookup to American's central computer in McGaw Park, Ill. Incoming orders are immediately routed by this computer

FIGURE 2.3
Chapter Summary

System Types and Functions	Examples	Key Design Features	Related Organizational Features
Operational: to structure work	Letter of credit Computer-integrated manufacturing	Work rationalization Work routinization	Work force composition Job design Organizational structure, work flow coordination Organizational culture
Monitoring and control: to evaluate performance and motivate people	Space management Typing productivity measurement	Standards Measures Evaluation Feedback Reward	Job design Organizational culture
Planning and decision: to support intellectual processes	Planning models Decision support systems Expert systems	Models Data manipulation	Work force composition Job design Organizational structure Organizational culture, centralization versus decentralization
Communication: to augment human communication	Message systems Teleconferencing	Communication procedures Communication mediation	Spatial and temporal factors Communication channels and networks
Interorganizational: to facilitate interorganizational transactions	Cash management for corporate banking customers Wholesale distributors' order-entry systems	Procedures for interorganizational transactions Mediation of interorganizational transactions	Relations with customers or suppliers Relations with competitors

to the company's regional distribution center that is located nearest the customer. American now has 90 of the regional centers scattered around the nation. Within minutes of typing in a purchase order on its terminal, a hospital will get back from American a printed confirmation giving the price of the items ordered and their delivery date. American claims that this system, called ASAP, for automated systems analytical purchasing, will cut the cost of a typical purchase by 20%. Hospitals can also save money by reducing their inventory levels, because American says that it ships 95% of all orders on the same day that they are received.

By combining the ASAP system with American's 18-month-old inventory control system, a hospital can track incoming supplies, disbursements, and stock levels and compare the projected inventory with actual supplies on hand to determine the best inventory levels of each department. The on-line, time-sharing system is augmented by a management service from American that tailors the program for each hospital. ["Systems that Slash Costs," 1980: 76E]

CONCLUSION

This chapter explored five types of systems, classified by the function each serves in the organization. The types are independent of technology: Unautomated instances of each type can be found, as well as examples that utilize a wide range of technologies. Similarly, the types are independent of users. Any system may have multiple users from different occupations or hierarchical levels. Each system type can be identified by means of its characteristic design features, which relate to observable organizational features. Figure 2.3 summarizes the function, design features and related organizational features of each system type.

The related organizational features are important because they are clues to the likely system impacts on organizations. System impacts are discussed in detail in chapter 3. Some systems may not fit neatly into the classifications presented in this chapter; they may have subsystems of different types. In cases like these, more than one set of organizational features should be considered in order to predict likely system impacts or to explain the impacts that have occurred.

The Impacts of Systems

INTRODUCTION

Chapter 2 described the packages of design features that characterize different types of systems and discussed the organizational features with which they are related. This framework can be used to identify the areas in which systems have had or are likely to have impacts. For example, when the design features of operational systems interact with the feature of work force composition, impacts are likely to occur in the areas of job opportunities and career prospects. Each organizational feature related to a system design feature is an arena in which organizational impacts are probable.

This chapter works through the design features framework, identifying the areas of impact and illustrating them with examples. The examples were chosen primarily to illustrate the impacts as graphically as possible and secondarily to present a balanced picture of both negative and positive impacts. Negative impacts may appear to have been given greater weight than positive, however, for two reasons. First, negative impacts are problematic; positive impacts are not. People want to do something about negative impacts to remedy them when they occur and to prevent them before they occur. Second, few impacts can be unambiguously classified as positive or negative. An impact that appears beneficial from one standpoint may appear harmful from another; a system that achieves the objectives of one individual or group may have costs for another.

It is important to realize that the impacts discussed here are possible but not inevitable. The documented experience of isolated cases and the results of systematic research lead to the conclusion that impacts of any type occur in some cases but not in others. Rather than invalidating the design features framework, this conclusion makes the framework more useful and necessary. If systems inevitably had the beneficial impact that designers intended when setting out to build them, there would be no need to think about negative impacts or, indeed, for this book. If systems inevitably had negative impacts, the only courses of remedial or preventive action would be to disband them or not to build them in the first place. Because this book is based on the assumption that the negative organizational impacts of systems can be predicted and controlled, a framework for identifying their likely, but not inevitable, impacts is indispensable.

OPERATIONAL SYSTEMS

Operational systems, like Citibank's letter of credit system (chapter 2), rationalize and routinize work. In so doing, they interact with the organizational features of work force composition, job design, organizational structure, work flow coordination, and organizational culture. These interactions can result in corresponding impacts on job opportunities and career prospects, job content and job satisfaction, the horizontal dimension of organizational structure, and social interaction patterns. Each of these impacts, summarized in figure 3.1, is described and illustrated in the following sections.

Job Opportunities and Career Prospects

When operational systems are designed and installed, organizations expect to derive benefits from streamlining their operations. Profit-making organizations expect lower costs from reduced utilization of raw materials and labor. Nonprofit organizations expect to achieve their objectives more fully with existing resources. However, the individual, as well as the organization, is affected by these changes. Sometimes the changes result in benefits to the individual, by expanding employment or career opportunities. Other times, the changes restrict or eliminate the individual's prospects in both the short and the long term.

Some evidence supports the argument that the impacts of operational systems are beneficial to individual workers. Computer-based systems may offer employment opportunities that would otherwise not exist for these individuals. At the same time that one newspaper article proclaims that many

FIGURE 3.1
Interorganizational System Impacts

Related Operational Features	Impact	Nature of Impact
Work force composition	Job opportunities	Expansion versus contraction
Job design	Job content and job satisfaction	Enrichment versus deskilling
Organizational structure, work flow coordination	Horizontal structure	Integration versus differentiation
Organizational culture	Social interaction patterns	Involvement versus isolation

jobs on the assembly line are lost forever "as Detroit prepares to follow Japan's lead to robots" (Nyhan, 1982), another describes how the use of robots on the assembly line allows Japanese firms to hire handicapped workers because the robots can hand parts to the workers (Lehner, 1982). In the United States, computers and telecommunications have enabled some innovative companies to employ programmers and word-processing typists to work at home (Mertes, 1981; Olson, 1983). Some of these workers are handicapped; others are parents who are unable to accept employment outside the home, because they have small children.

Labor cost reductions due to systems may allow firms to reduce prices and increase their volume of business. While fewer workers are needed per unit of work, more work is done, leading to an increase in number of workers. Sometimes, systems create more jobs than they eliminate, a point made by a sardonic cartoon in the *Wall Street Journal* (February 24, 1982): "Harris, you're being replaced by a computer, three programmers, two repairmen, two computer operators, six. . . ." In support of these arguments, a 1981 U.S. Department of Labor study concluded that computer-related job displacement over the past thirty years "has been more than offset by jobs created in new industries and by heightened productivity in established economic sectors" (Kirchner, 1981). A newspaper article quoted from the report that:

> [A] steady stream of technological progress . . . has resulted in higher productivity, elimination of many menial and dangerous jobs, higher wages

and shorter hours and a continuous flow of new products and services, which have resulted in a higher standard of living.

Finally, many organizations have retrained or found new jobs for workers displaced by systems. Time freed up from repetitive or dangerous jobs has been made available for planning and problem-solving activities that were previously neglected. In addition, by waiting for normal attrition through promotions, retirements and voluntary quitting, firms have avoided firing workers whose jobs are made redundant by technology.

Critics of these optimistic views abound, however. The newspaper article covering the Labor Department study (Kirchner, 1981) notes that the report "did concede . . . that technological innovation has eliminated jobs in some industries and caused a certain amount of 'painful' employment adjustment."

Kelly Gotleib, a noted Canadian computer scientist, concludes from the evidence of dozens of studies that jobs lost through computing cannot be offset by jobs gained, largely because the new jobs are not filled by the displaced workers (French, 1980).

Concerns about the impacts of computers on employment run higher in Europe than in the United States. One major concern of European labor unions has been what is called "silent firing," the increase in productivity that enables employers to avoid hiring in the future. One French study estimated that banks in France would be able to provide better service while hiring 30 percent fewer workers over the next few years because of computer-based technology (Nora and Minc, 1979). A study in Norwegian banks found that the use of computer terminals in branches enabled a 50 percent increase in the number of transactions handled, with no increase in employment (Norsk Regnesentral, 1979). In Sweden, labor unions have become so concerned about job elimination and the decreased need for skills (discussed in the next section) that they have proposed a moratorium on the introduction of new technology until studies of technological impacts on jobs can be done (Chamot and Dymmel, 1981).

Banking is not the only industry in which such concerns have been raised. A few years ago, air traffic controllers received national attention for an illegal strike to protest the detrimental health effects of their high stress jobs, entailing constant use of video-display terminals (VDTs). A number of newspaper articles in the wake of the unsuccessful strike have suggested that increased air traffic control automation would eliminate similar problems in the future:

> In the next decade, FAA planners say, the number of controllers could be more than halved by letting computers take over many of the time-consuming jobs humans now do.

AERA (automated en route air traffic control system) was being developed even before the nation's air traffic controllers went on strike. But it has been discussed as a solution of the FAA's labor problems since the first rumblings of a controller walkout. [Lattin, 1981]

Job Content and Job Satisfaction

The rationalizing and routinizing functions of operational systems may affect the design of jobs and workers' reactions to them. On the one hand, operational systems may enrich jobs by expanding the importance, variety and coherence of job tasks. Jobs with these attributes are believed to create high job satisfaction. On the other hand, operational systems may reduce the need for skills to perform jobs, depriving employees of the creative, thoughtful tasks, leaving the boring and repetitive activities that breed discontent or apathy.

Mattheis, writing about Citibank's letter of credit system, claims that the system enriched jobs and generated job satisfaction:

> This shift in job content and environment has had enormous impact on our employees. They are no longer cogs in an assembly-line wheel; they are providers of services to real customers. Gone are the stultifying effects on human energy and motivation often caused by the specialization of labor. People have full responsibility for an identifiable service that they personally deliver to individual, known customers.
>
> The work-station professionals have responded visibly. They take a proprietary interest in both their customers and the knowledge they have about those customers, the transactions, and the technology they 'own.' One man placed the flags of all the countries he serves atop his CRT terminal. People tend to feel displaced and to get edgy if someone else needs to use 'their' equipment for a minute. The employees are talking to customers again, as in the old days. They are answering questions and handling problems about transactions they themselves have processed, getting instant feedback on their own performance from the people they perform it for. . . .
>
> Those workers . . . have in fact moved from single-function clerical jobs to skilled, professional assignments with a corresponding upgrading of compensation. People who had been hired as clerks to perform a single task for which they received weekly wages are today regarded by this corporation as professionals in whom the corporation has happily made an important investment.
>
> Many of these former clerks are now officers of the bank; all of the jobs in the letter of credit department today have the potential to lead to officer status, and formalized career paths exist for all personnel in the area.

The salary range of these professionals is at least double what the salary wage was when they were clerks, and the potential for growth is, of course, very great. [1979: 157–158]

In contrast, the following excerpt describes a job that has been stripped of skill through the introduction of a computer-based system:

COMAS [Computerized Maintenance Administration System] stores information on past malfunctions in switching equipment and is used by technicians as a handy reference system to help identify the sources and causes of new problems. By comparing present difficulties with past performance, the job of tracking down breakdowns is simplified. Before COMAS, foremen were never directly involved in the work of tracing and repairing switching equipment malfunctions. But when the system was introduced in 1971, management personnel began using it to do what had always been the work of switching technicians. . . .

In effect, the job of the switching technicians is being split in two. The most interesting and skilled tasks are disappearing into management, leaving behind tasks shorn of skill and reduced to clerical work. [Howard, 1980: 24]

While many people have analyzed and described the process of eliminating skills from professionals' and craftspeople's jobs (e.g., Braverman, 1974; Kraft, 1977), few have conducted systematic studies on the enrichment versus deskilling issue. Kling's study (1978a) is a refreshing exception. Kling tested the effects of computing on jobs and work through a study of 1,200 managers, data analysts and clerks in forty-two municipal governments. His findings supported the hypothesis that computing increases rather than decreases skills needed to do jobs, but he also found that these beneficial impacts were stronger for managers than for clerks. Kling's results indicate that the effects of computing on job design are relatively benign on the average. As with the findings of the Labor Department study (Kirchner, 1981) on employment, however, this average may conceal some cases of extremely detrimental impacts.

Horizontal Structure

Operational systems rationalize and routinize work, in part, by changing the way that the work flow among jobs and departments is coordinated. Consequently, it seems reasonable to suspect that changes in the horizontal or lateral dimensions of organizational structure are caused by systems.

Unfortunately, not much research evidence supports or refutes this hypothesis. Relatively little conceptual or empirical work has been done on the role of systems in cross-departmental relationships. For example, while Galbraith (1977) describes the role of information systems in the vertical (hierarchical) dimension of structure, he ignores the role of systems in the lateral relations among departments and other units. Robey (1981) reviewed the research in this area and drew one unambiguous conclusion: The greater the horizontal differentiation or complexity of organizations, the greater the use of computer-based systems. While this could mean that systems make structures more complex, it is more likely to mean that more complex structures require systems for work flow coordination. Therefore, a change in formal structure may be the impetus for the design of a new system for coordination. In contrast, a new operational system may make an altered organizational structure possible or necessary. These interactions between systems and structures are clearly visible in the history of Citibank's letter of credit system.

Citibank's management began to perceive operational problems in the bank around 1970. At that time, they understood banking operations as a huge pipeline of transactions. Organizational subunits (mail room, correspondent bank section, etc.) and managerial responsibility cut across the pipeline. Every downstream operation depended upon the activities of the upstream operations. This dependence made downstream units relatively powerless to affect the processing of particular transactions.

Citibank's evolution toward current operational systems entailed two major changes in the way management conceptualized the work flow. The first replaced the view of the bank as a pipeline of transactions with the view of the bank as a factory.

> In response to this problem, the management team instituted a production management tool, the assembly line. It began by breaking out the items by source. . . . Independent processing organizations were established for each of these major distinctions.

> Within these, refinements were made according to transaction type, that is, by product line. A separate 'channel' was established for each product, and all the functions needed to process that product were included in the channel. We had then complete vertical management: a person who managed a channel now had control over the total transaction from the time it entered the bank until the customer was advised of its completion. . . .

> The restructuring itself was based on a fundamental conceptual difference: that the back office was really a factory, not a clerical operation. From this starting point, the techniques of production management were adapted to the special realities of our products and to the kind of work involved in producing them. [Mattheis, 1979: 148–149]

Several years later, about 1975, Citibank reconceptualized its operations. Here the emphasis was shifted away from factorylike processing efficiency toward service-oriented customer relations:

> A customer buying a product off a shelf chooses one of many identical products. A customer buying a service is asking for a specific activity that will meet a specific need. Clearly, we would have to go beyond production management solutions to a post-industrial model of services management. This added dimension required stretching our vision to look outside our processing shop to where the customer sat, and then to look back in from his viewpoint. . . .
>
> For the customer, after all, the window into his bank should be a single entry point, angled toward his marketplace. The actual processing of his transaction should be invisible to him, but fast and accurate. The service he is provided should come from a responsive human being who, through familiarity with the customer and his business, becomes more than just a disembodied voice on a telephone or a number on an inquiry form. [Mattheis, 1979: 151]

It would be difficult to justify the claim that the introduction of new operational systems caused the changes Mattheis described in Citibank's organizational structure. Mattheis argued that structural changes were introduced simultaneously with systems to achieve the desired redirection of the bank's efforts:

> We devised a new program that has three basic components: first, a reorganization along market-segment lines; second, the establishment of the appropriate technological base to serve the production needs of this market-driven structure; third, the redesign of jobs and processes. [1979: 151]

Nevertheless, the two types of changes served to reinforce each other. The need or desire for the structural change was also the need or desire for the operational systems, and vice versa. In this sense, a change in organizational structure may be described as an impact of operational systems.

Social Interaction Patterns

Changes in job design and organizational structure often alter the patterns of interaction among people in a firm. Therefore, operational systems may have concomitant impacts on social interaction patterns. By altering job design and by restructuring organizations, operational systems may isolate people and departments that were previously central in social interaction or in the work flow, and vice versa.

Many people have expressed the concern that computerized systems produce jobs in which people remain socially isolated from colleagues while tucked away at their "neat, cockpit-like workstations." Some of this concern can be seen in a quote from the Citibank article:

> Our success with the work station has not, however, been unqualified. Through the regular feedback sessions that Citibank arranges for employees, and through supplemental interviews specifically with the letter of credit work force, we have learned that the single-product work station may have swung the pendulum a little too far back. While the workers like the responsibility and the content of their "beginning-to-end" jobs, a certain feeling of isolation is also present. The workers still help one another—in a kind of ongoing training—and they are able to socialize freely and well. What is lacking, however, is a real sense of teamwork, as well as the support and greater confidence that teamwork can instill. [Mattheis, 1979: 158–159]

Similarly, departments may experience increased or decreased centrality in the work flow because an operational system rearranges the flow of information or of materials. This can result in changes in the distribution of power within the organization, a topic that is explored more fully in later sections of this chapter.

MONITORING AND CONTROL SYSTEMS

Monitoring and control systems, like FBT's Space Management System, track and evaluate people's performance. They interact with organizational features of job design and culture, resulting in impacts on autonomy and control, which is strongly related to job satisfaction; organizational psychology and organizational performance. Each of these areas is discussed in detail in the next sections. Figure 3.2 summarizes the impacts of monitoring and control systems.

Autonomy and Control

Autonomy is a characteristic of jobs in which the worker has the freedom or discretion to schedule the work or to choose work methods. The greater the supervision of a worker's behavior, the less the autonomy in that worker's job.

FIGURE 3.2
Monitoring and Control System Impacts

Related Operational Features	Impact	Nature of Impact
Job design	Autonomy and control Organizational psychology	Self-control versus external control Well-being versus powerlessness and stress
Organizational culture	Organizational performance	Desirable organizational behavior versus game playing and abuse of systems

Autonomy is believed to be one of the most important ingredients of job satisfaction, along with skill variety, in which the job requires the exercise of several different activities and skills; task identity, in which the job entails the completion of an identifiably whole piece of work (rather than a specialized subtask like turning screws all day long); task significance, in which the job is believed important by the worker, and feedback from the job, in which the job gives the worker information about results that can be used to improve performance (Hackman and Lawler, 1971; Hackman and Oldham, 1975).

Monitoring and control systems are believed by some individuals to increase autonomy; others claim that these systems can reduce autonomy and increase the degree of supervision or external control over the worker. The argument in favor of increased autonomy runs like this: Systems can permit greater delegation of responsibility to lower-level managers and workers because the system provides them with the information they need to make decisions and because the system quickly and automatically evaluates their performance. The latter feature ensures that the errors of lower-level employees will be caught before they have disastrous consequences (Pfeffer, 1978).

In contrast, the system can be used to monitor closely not only the results of actions but also the actions taken; this feature enables supervisors to enforce prescribed ways of doing things. Howard describes this process as it relates to telephone operators at AT&T.

> Working a TSPS [Traffic Service Position System] terminal is easier than the
> old cord-board. . . . This does not mean that operators have more control

over their work; in fact, they have considerably less. With the cord-board, operators could regulate somewhat the pace at which they responded to calls. . . . [The new system] means an operator can handle an unending succession of calls. There is no such thing as a full terminal. . . .

For half an hour, two times each week, every operator is timed by computer to determine her 'average working time' or AWT. . . . Operators are evaluated on their 'speed of answer'. After the electronic beep, they have three seconds to respond to a call. . . . Many still complain that the pace of their work has increased. . . .

In addition to the twice weekly AWT productivity studies, each operator is monitored by a supervisor for thirty calls per month to evaluate courtesy and accuracy. . . . Like the AWT studies, operators do not know when they are being observed. . . .

There are entirely separate offices which listen in on TSPS groups, evaluate their work, and inform their supervisors. . . . The computerization of work has intensified this sense of powerlessness. . . . [1980: 26–27]

Organizational Psychology

The last sentence in the preceding excerpt illustrates an area of impact closely related to impacts on job characteristics and satisfaction—namely, the psychological adjustment of workers to their jobs and work environment. Some people claim that systems have beneficial psychological effects by eliminating repetitive and boring work and by giving workers the tools to improve their performance on the job. Others fear that systems can decrease psychological well-being, leading to increased stress and possibly even to deterioration in physical health.

Argyris (1970; 1971) has argued that rational management information systems can create serious emotional problems among the executives who use them. Among these are four negative psychological reactions to the systems that make managers' behavior more visible and subject to external control:

1. Reduction in psychological freedom of action,
2. Feelings of psychological failure,
3. Undermining of formal authority and political relationships because of emphasis on conceptual thinking and demonstrated competence,
4. Lessened feelings of powerfulness and importance.

Argyris points out that some information systems produce reports indicating a desirable course of action. One consequence of this is for managers to feel that fewer options remain open to them. Options that might have involved politics and interpersonal influence are most likely to be excluded from a manager's behavioral repertory. Because of these restrictions, the manager may come to feel hemmed in and powerless.

Powerlessness may result not only from the tendency of systems to restrict actions but also from the fact that systems may embody goals and standards set by others. Individuals motivated to achieve and to behave responsibly may feel they have failed, even when they perform well, if the goals and performance standards were set by others.

Further, systems are designed around hard, quantifiable performance data; softer, less tangible data rarely figure in systems logic. Consequently, Argyris argues, systems focus greater attention on the qualities that are measured. An ability to make the numbers come out right may come to be valued more than formal authority or political skills. While many people believe that leadership should be based on competence, not on position or influence, Argyris points out that a change in the nature of the organizational game can be devastating to those who used to play well by the old rules.

Finally, managers derive positive feelings of self-worth from facing ambiguous situations with a decisive course of action. In situations like these, executives can believe that their actions really make the difference. However, when the computer deciphers a situation and proposes a plan of attack, a manager may well feel that his or her function has been reduced to that of custodian to a machine.

Zuboff (1982) has argued that systems may increase the control of managers over workers while they hide the true source of the control. Systems may replace direct, personal supervision with invisible electronic supervision. How much more powerless and vulnerable will people feel when they cannot even identify who or what is telling them what to do? What happens when the supervisor is in the machine?

Garson (1981) illustrates the point clearly:

> Tuesday, 10:30 P.M.: The lone computer operator comes over to my console and says in a friendly way, "If you're going to stay here, you'll have to get your productivity up."
>
> "Oh," I say, "what is my speed and what should it be?"
>
> "It's been scientifically set," he tells me, "at 50,000 keystrokes an hour."
>
> Then he sits down, plays a couple of chords on his control panel and up come my figures. The figures show when I started—to the nearest tenth of a second—when I took a break and exactly how many keystrokes I'd done all evening. (I am very far below the 50,000 keystrokes an hour.) The real supervisor is inside the machine.

Feelings of powerlessness and invisible external control may lead to increased stress and, ultimately, to deterioration in physical health. A growing body of research supports concerns about the safety and health of workers who spend their days in front of VDTs. Complaints of fatigue and strain to eyes, necks and backs are common, and many people worry about radiation from the ray tubes. A National Institute for Occupational Safety and Health (NIOSH) study found that VDT operators had higher levels of anxiety and irritability and more eye and muscular complaints than workers who did not use terminals all day (Lublin, 1980). According to another source, the level of stress found among the VDT workers was the highest of any group studied by NIOSH, including air traffic controllers (Gregory and Nussbaum, 1981).

Organizational Performance

A third area of monitoring and control system impact is that of organizational performance: What people do with systems and how well they achieve the results desired by the organization. Many organizations have installed monitoring and control systems and have found, first, that people used these systems as their designers had hoped and, second, that this use resulted in improvements on performance dimensions important to the organization. Others have found, however, that people use systems in ways other than those intended—e.g., by obeying the letter of the system but not its spirit—and that the result is a failure to achieve the desired performance level or a deterioration of performance in other areas.

Consider FBT's space management system. The system was used originally to monitor space utilization and to ensure that GSA space charges to FBT were accurate. However, FBT's national office soon began using the space utilization rankings as a criterion in the performance evaluation of the regional and district administrators. In 1976, for example, the space program manager announced that the FBT was not going to reward regions that have too much space by giving them more money than the other regions. Budget cuts would come first from the regions with the highest utilization rates.

Within about six months, problems with SMIRS began to surface. Some districts and regions began to play games with the system, timing their data entry to present the most advantageous picture of their space utilization. Space assignments in areas with high utilization rates were shifted to areas with low rates. Staffing numbers were gradually inflated to decrease the ratio of space per staff. This practice was apparently so widespread within FBT (without any overt collusion among regions and districts) that, within two years of introducing SMIRS, the system reported a total FBT staff that was greater by 10,000 people than the official payroll.

In addition, many units within FBT began lobbying efforts to discredit the system or to change it. Most likely to protest were those units high on the ranking by space utilization rate. These units generated various proposals to have the system measure space cost per employee rather than square footage per employee or to count authorized (but unhired) personnel in addition to those already employed. Some districts simply refused to enter their data into the system.

FBT ultimately corrected the problems with SMIRS by adjusting the measures of space utilization and the way in which they were used in performance evaluation. In some situations, however, the negative aspects of a monitoring and control system may avoid detection for a long time and may result in irreparable damage. Consider the scandal in the advertising firm of J. Walter Thompson. In February 1982, the firm announced:

> [I]t would charge $18 million against JWT's pretax earnings for the period 1978 through Sept. 30, 1981. Two days later, the company said its 1981 fourth quarter earnings would be reduced by a further charge of $6.5 million before taxes. JWT explained that the revenue of Mrs. Luisi's unit, which had been reported as totaling $29.3 million during the 1978–81 period, included $24.5 million that consisted of "fictitious entries" in the unit's computer.
>
> JWT officials haven't accused Mrs. Luisi herself of wrongdoing. . . .
>
> The bookkeeping scheme—whoever perpetrated it—started small in 1978, when Mrs. Luisi's unit evidently fell a bit short of its revenue and profit targets, JWT officials say. Phony computer entries made it appear that the unit had met its goals, but this, in turn, led to the unit's targets being set higher for 1979. When the 1979 targets weren't met, more bogus revenue was booked, leading to a rise in the *next* year's targets—and the cycle continued on a vastly increased scale in 1980 and 1981, according to the company's officials. [Blustein, 1982]

The article goes on to attribute the misuse of this system to faulty internal controls and reporting procedures coupled with intense pressure to perform. These same factors weighed heavily in the Equity Funding scandal and other similar cases of computer fraud inside organizations. The problems of the FBT may appear to belong to a totally different class than the phenomenon of crime by computer, but they both have their origins in the same kinds of organizational processes.

PLANNING AND DECISION SYSTEMS

Planning and decision systems such as planning models, decision support systems and expert systems rationalize and routinize the intellectual aspects of work. They interact with the features of work force composition, job design, the centralization versus decentralization of organizational structure and organizational culture, with resulting impacts on job opportunities and career prospects, job content and job satisfaction, decision making and power and politics. Figure 3.3 summarizes these interactions and the nature of the resulting impacts.

The impacts of planning and decision systems on job opportunities and career prospects and on job content and job satisfaction are quite similar to those of operational systems on these same dimensions. Consequently, these dimensions do not receive separate treatment here. It is important to note, however, that planning and decision systems are used more extensively by managers and professionals than by the blue collar and clerical workers who are assumed to be the usual victims of contraction in career prospects and the elimination of job skills.

Since the 1958 publication of Leavitt and Whisler's now classic article "Management in the 1980's," people have speculated that the computer will induce changes in managerial employment and career prospects similar to those observed in certain lower-level occupations. The authors predicted that computerization would automate the jobs of many middle managers and would reduce the prospects of those remaining for promotion to top management. Leavitt and Whisler also predicted that the use of computers in

FIGURE 3.3
Planning and Decision System Impacts

Related Operational Features	Impact	Nature of Impact
Work force composition	Job opportunities and career prospects	Expansion versus contraction
Job design	Job content and job satisfaction	Enrichment versus deskilling
Organizational structure, centralization versus decentralization	Decision making	Increased centralization versus decentralization of decision making
Organizational culture	Power structure	Shift in balance of power
	Politics	Increased versus decreased political behavior

managerial decision making would sharpen the distinction between planning and doing, that planning would be forced higher up in the organization and that middle managers would be left with jobs bereft of decision making. Until recently, very few high-level managers have used computing in a direct hands-on way, and investigators have produced little evidence to support this contention. Within the last few years, however, increasing concerns about the elimination of management skills have surfaced. A special report in *Business Week* heralded "A New Era for Management":

> As the electronics revolution takes hold and offices and factories become computerized, data can flow directly from the shop floor to the executive suite, making many middle managers redundant. Those who survive will have far different jobs to do; those who don't will be hard put to maintain their middle-class existence. [1983: 1]

The remainder of this section focuses on two areas of impact not covered in the section on operational systems—namely, impacts on decision making and on power structure and politics.

Decision Making

One important metric of organizational structure is centralization, variously defined as the level in the hierarchy at which decisions are made or the degree to which decisions are made at headquarters rather than in the field. Organizations with a divisional structure are believed to be more decentralized than those with a functional structure (see figure 3.4). Degrees of centralization exist, however, even within so-called decentralized firms.

Leavitt and Whisler (1958) predicted that computers would cause decentralized organizations to recentralize. Their argument went like this. After World War I, many industrial organizations quickly grew very large. Prior to this growth, these firms may have had only a single product line and may have been local or regional in geographic scope. Under these conditions, it was easy for a small number of decision makers to understand and manage the firm's operations. The simple functional division of labor was common, and managers at the top coordinated and integrated the various functions.

Growth in size and diversity created difficulties with this arrangement. As firms entered new markets, they encountered new problems, more and more of which had to be referred up the hierarchy for resolution. Top decision makers were swamped with requests for answers. To lessen their overload, they decentralized their firms by creating divisions for different product lines, assigning some members of each functional specialty to each division and delegating to each division head much of the responsibility for decision making about their products.

FIGURE 3.4
Organizational Structures

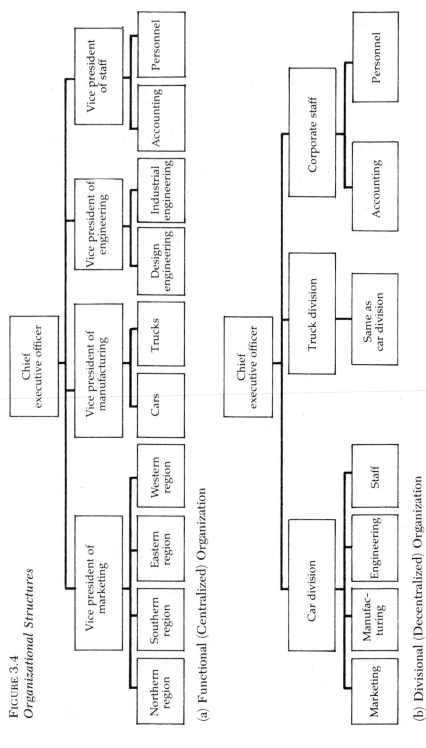

(a) Functional (Centralized) Organization

(b) Divisional (Decentralized) Organization

Leavitt and Whisler argued that top managers decentralized only reluctantly; they would have preferred to maintain control if there had been a way to do so. At that time, there was no other way, but the introduction of computing provided an alternative. With the fast and accurate reporting of results in computerized information systems, top managers had a means to manage information overload while they retained their ability to make all of the key decisions. Leavitt and Whisler concluded that top managers would seize this opportunity. Consequently, the locus of decision making would move up the hierarchy, and structures would shift from decentralized to centralized.

This reasoning generated some heated debate. A few critics argued that top managers were content to do long-range planning and policymaking and to leave operations to their subordinates (Burlingame, 1961). Some theorists constructed an equally plausible rationale that computerization increases delegation and decentralization (Pfeffer, 1978). They argued that higher-level managers are reluctant to delegate decision-making tasks to lower-level managers because of the long time between decision making and knowledge of the results. If an initial decision is bad, but if one does not find out for six months that it is bad, the decision maker will probably continue to make bad decisions for six months. Under such conditions, a superior has limited ability to monitor and correct poor decision making by subordinates and, thus, is unlikely to delegate.

With computerized information systems, however, more timely and accurate reporting of results reduces the negative consequences of bad decisions by subordinates; bosses can leave subordinates alone for a while, knowing that they will receive reports telling them how their juniors did before the damages can pile up. This ability should encourage managers to delegate more, not less. Of course, one could argue that delegation of decision-making authority, coupled with greater supervision and control, hardly represents true delegation or decentralization of decision making, but this does not weaken the logic of the argument.

The research does not lend unanimous support to either argument. Several studies have been conducted, with contradictory results. Robey (1976; 1977) reviewed these studies and suggested a number of factors that could account for different results in different firms. Subsequently, Robey (1981) helped design and conduct a multicountry study in which it was possible to make before-and-after comparisons of decision-making centralization in each firm due to system use. He found that:

· No change in structure occurred more frequently than change, whether toward centralization or decentralization;
· When the structure did not change (either from centralization to decentrali-

zation or vice versa) the original structure, whatever it was, was reinforced, not weakened;
· Where change did occur, the structure was more likely to change toward greater centralization;
· Greater centralization was accomplished by locating decisions lower in the organization but by increasing supervision and control over performance;
· In no case were computer systems used to encourage democratic or participatory decision making.

It seems that Leavitt and Whisler (1958) were closer to being right than wrong.

Power Structure and Politics

Power, like structure, is a central concept in many organizational theories. Power can be defined as the ability to get one's way in the face of resistance or opposition (Pfeffer, 1981). Politics is the process of exercising power. Power is usually considered to be an attribute of individuals, related to their ability to influence others to do what they want. In this view, an individual has power to the degree that he or she possesses certain things or characteristics that generate power, such as charisma, expert knowledge, resources or formally designated power (French and Raven, 1959).

Each subunit or department of an organization can also be described as more or less powerful. For example, one department can be said to be more powerful than others, if decisions about which disagreement exists tend to favor that department consistently over others. One well-supported explanation of this phenomenon is that certain subunits are abler than others to solve the strategic problems confronting the organization (Hickson et al., 1971; Hinings et al., 1974). Subunits can do this when they occupy a central position in the flow of work through the organization, when the problem they can solve is viewed as very pressing by people in the organization and when there is no easy substitute for the services they provide. In addition, organizations can gain power over individuals or other organizations by supplying resources on which others become dependent (Pfeffer and Salancik, 1978).

Power is closely interrelated with the concept of structure. Location within the formal hierarchy can provide some actors (people or units) with bases of power like the ability to make decisions affecting others, control over the resources on which others are dependent and access to the physical manifestations of power. Thus, people tend to equate power with legitimate authority. In fact, however, a vast literature amply demonstrates that people and groups can acquire and use power quite out of proportion to formal authority. To cite just a few examples, Crozier (1964) described the power of maintenance workers in a tobacco-manufacturing plant, Mechanic (1962) explained how lower-level hospital employees could exert power over high-status professionals and Strauss (1962) discussed the political tactics of a not-so-powerful group of purchasing agents. Therefore, in exploring the concept of power, it is useful to assume that power is distributed unequally in organizations, that this distribution may differ significantly from the formal organization chart but that the power distribution is nevertheless not random but related to observable organizational features.

The adage "Information is power" captures in one phrase the way in which planning and decision systems can change the organizational balance of power. Employees can acquire power by obtaining information that others need to do the things that are important to them. Actors can maintain their power by controlling access to this needed information; as long as there is no other source for it, others will need the data custodians. In addition, actors with control over a single source of needed data can exercise power by selecting and filtering what they mete out. Everyone knows that one can lie with statistics: How information is presented may strongly influence how it is interpreted. Consequently, actors who control information may be able to affect the result of a decision-making process in ways that favor them or their objectives.

Pettigrew (1973), for example, has described a case involving the decision to purchase a computer. A manager favored one computer manufacturer over the vendors that were chosen by his subordinates. Although the manager may have lacked some of the technical expertise of those reporting to him, he was able to swing the decision his way because he made the presentations about the various computer models and their capabilities to the people who made the final decision. Pfeffer (1981) gives many other examples of the control of information in the politics of decision making.

The introduction of new or altered information systems can disturb existing patterns of information flow within a company or a department. Any number of quite different changes can produce a disruptive effect. Thus, both centralizing previously decentralized data and decentralizing access to previously centralized data can create problems. Consider an example of a planning and decision system for logistics in a branch of the U.S. military (Conrath and du Roure, 1978).

The department in question was charged with keeping track of the location of equipment and with moving it to other locations upon appropriate request. The department was rigidly hierarchical; communication and directions tended to follow formal reporting relationships. For example, inquiries about the location of equipment or requests for transfer of equipment were initially directed to officers at a relatively high level who then requested subordinates to supply the appropriate details.

Before the advent of the new computer-based logistics system, data about equipment location were compiled laboriously by junior officers. These people had developed an elaborate network of telephone contacts who could tell them what they needed to know. The manually generated reports they eventually forwarded to higher management were about a month out of date on the average.

The new system gave each junior officer an on-line terminal with status information and programs to determine the least cost route to move equipment between locations. When the new system was installed, people started making requests for information and for equipment transfer directly to the junior officers with the terminals rather than to the appropriate senior officers. Although they were explicitly forbidden to authorize the equipment movement without the proper permission, the junior officers could not resist the temptation to do so because the system made it so easy. Thus, the system undermined traditional authority relations, and one would expect this to be unacceptable in a military environment. The senior officers vigorously protested against the system, claiming that the data it contained were inaccurate. The system was only saved by an expensive redesign effort that, it seems, disabled the capability of junior officers to use the system in ways that circumvented formal channels.

COMMUNICATION SYSTEMS

Communication systems such as message systems and teleconferencing systems augment human communication, affecting the organizational features of time and space and communication networks. This interaction results in impacts on location of work and geographic dispersion and who communicates to whom. These impacts, summarized in figure 3.5, are discussed in the following sections.

Location of Work and Geographic Dispersion

Many futurists have predicted that increased use of communication systems will create an era when many of us work at home, linked to other organizational members by terminals and other communication devices. Work at home is expected to eliminate lengthy, energy-consuming commutes and, incidentally, to lower companies' bills for rent, maintenance, utilities and furniture (Toffler, 1980; Martin, 1981). Others have questioned the benefits of working at home.

FIGURE 3.5
Communication System Impacts

Related Operational Features	Impact	Nature of Impact
Spatial and temporal factors	Location of work	Traditional work place versus work at home (telecommute)
	Geographic dispersion	Increased or decreased organizational size and/or number of locations
Communication channels and networks	Who communicates to whom	Efficiency changes versus network changes

People working at home might miss the valuable social contact and socialization processes that occur in offices. Supervision patterns will change, and supervisors will probably find the adjustment difficult. In addition, while some workers will value increased time with their families, others may find constant contact stressful and disruptive. Working at home may disorient people who have relied upon the physical separation of settings for clues about appropriate roles and behaviors (Becker, 1981; Olson, 1983).

Becker and McClintock (1981) report preliminary results of studies of stockbrokers, computer programmers and Tupperware dealers working at home. Among other things, they found that:

· Most stockbrokers already worked at home several hours per day and did not want to increase their time away from the office.
· Tupperware dealers wanted to work at home to be with their children more but saw them less when they did.
· Stockbrokers did not mind background noise from TV and machines but were distracted by sounds from family members.
· People working at home rarely had the space or the funds to set up a work space that was private and workable.

Olson (1983) reported on experimental programs which allowed certain workers to work at home. These programs were considered positive both by managers and workers. Olson raised a number of concerns independent of managerial issues such as what kind of work can best be done at home by what kinds of workers how supervised. These concerns centered on individual stress, the employee's relationship to professional peers, long-term career paths, organizational commitment, family relationships and social isolation.

The possibility of widespread working at home suggests the possibility of increasing organizational dispersion—that is, widening the geographic area over which organizational units are scattered. Historically, communication and transportation systems such as rail, telegraph and telephone have played an important role in the evolution of the modern firm from a small, single location enterprise to a large multisite corporation (Chandler, 1977). New communication systems, enabling many employees to work at home, could intensify these trends. The number of geographic locations operated by a company could become as large as the number of employees. This may create problems for companies that do not, cannot and/or should not exert extensive control over the homes of employees.

However, without the ability to supervise and control their workers, organizations may be required to find distinctly different modes of operation. One such mode is contracting out, where individual entrepreneurs, no longer employees, perform work for a fee. Some writers (Handy, 1980; Schumacher, 1973; Macrae, 1976) argue that this type of arrangement is a natural solution to the problems that plague large bureaucratic organizations; but it may lead to dismantling them.

Increasing geographic dispersion of organizations could raise substantial issues long before firms are dismantled, however. Some firms will undoubtedly use communication technology in lieu of making more appropriate structural and/or geographic changes. The systems may aggravate or at least fail to solve existing problems. Consider this example: Several large organizations have decided that their first use of video teleconferencing should be to link computer system developers more closely to those for whom they are developing systems. They see this as a solution to the problems of poor communication between users and developers. These problems were created largely by two prior organizational arrangements whose wisdom is not questioned: The decision to have system developers report organizationally to a central computer group instead of to the users of systems and the decision to locate the central computer group in its own building, at some distance from the users. (Not all organizations make these decisions, but many do. The consequences of these decisions are discussed later in chapter 7.) While teleconferencing may improve the symptoms created by these arrangements, it does little to address their causes. And while it is impossible to predict the net result, it is at least possible that the effects will not be entirely positive.

Who Communicates to Whom

Communication systems are claimed to reduce the constraints of time and space and to increase the efficiency of the communication process (that is, its speed, accuracy and completion). Communication systems may also change the nature of communication, however, not just its efficiency, by altering who communicates to whom about what, in addition to how fast and accurately they communicate. If so, the organizational impacts may be significant.

The possibility that systems may change the nature of communication is suggested by the following line of reasoning. If every communication over a new system replaces one over an older system without increasing the number of communications or communicators, one would expect only to observe changes in communication efficiency. If, however, the use of a new medium allows or encourages new communication, one would also expect to see changes in the network of communicators. Over time, changes in communication networks could lead to changes in the formal distribution of authority or structure and in the informal power distribution.

Some evidence supports this argument. For example, Palme (1981) studied a computer-based message and conferencing system in the Swedish National Defense Institute. Users of this system reported that 50 percent of their system messages replaced written memos, telephone calls and face-to-face meetings. They reported that the remaining 50 percent were new communications that would not have taken place without the system.

This system has two separate modes of use: One resembles ordinary mail, and the other brings into contact people interested in a particular subject area. Unlike the regular mail mode, the conferencing mode does not require a sender to know the names and addresses of the parties with whom he or she corresponds. Palme (1981) counted the number of communications of each type and prepared distributions according to the locations of sender/receiver pairs (same department, different department, different organization). The distribution for mail-type messages was significantly different than the distribution of conference entries, implying that communicators are reaching a very different group of recipients through the conferencing mode.

How might this lead to changes in formal authority structures or in informal power distributions? A strong relationship exists between the formal structure of an organization and actual patterns of communication. The correspondence between structure and communication is not perfect because of factors such as space and geography, interpersonal attraction and task flows that cut across structural units. However, formal structure both requires interaction among related parties and provides opportunities for interaction, for example, through the physical co-location of people in a department. In addition, many structural changes probably occur when cross-unit communication volumes exceed those inside the units. Therefore, changes in communication patterns can put pressures for changes on the formal structure.

Let's consider the evidence to support this contention. Conrath (1978) has also reported distributions of communication according to the organizational location of senders and receivers. He found significant differences across firms according to their organizational culture. Palme's distribution of communication via the mail mode resembles the distribution Conrath found in a bureaucratic organization (a branch of a public utility). In contrast, the distribution of conferencing communication distribution resembles Conrath's data from less bureaucratic firms (a manufacturing firm and an insurance company). This suggests that the use of a medium that engenders new communication rather than merely substitutes for older media may break down hierarchical authority relationships and cultural patterns.

An article in a computer trade journal provides an example of the process (Emmett, 1981). The system in question is a message system called VNET. VNET is a network connecting 400 computer central-processing units. IBM employees who use terminals in the course of their work can send messages to other terminal users over the VNET. The electronic mail capabilities of VNET evolved almost accidentally, since IBM management did not plan or sponsor it, but today "VNET could already be the world's largest electronic newspaper" (Emmett, 1981: 50). The designers of VNET were working on a project that ran counter to the strategies pursued by IBM management. The people on the project believed strongly that it was useful and persisted in working on it after management disbanded the group and scattered them to different geographical and organizational locations. VNET helped them stay in touch with each other, and they eventually succeeded in convincing management that they were correct by interesting several customers in the product idea. Management reluctantly allowed the product to be sold and earmarked funds for the necessary ongoing development, but it takes little imagination to guess that the architects of the coup made themselves something less than popular with the authorities.

As a result, people working on the project have had fears about their career prospects and many of them have left to join other firms. As concerns within this project group grew, VNET once again figured heavily, as a way for people to exchange gripes about management and about their uncertain situation.

> VNET has been used, among other things, for personal attacks on IBM management, to send job resumes, and even to announce resignations, sources say. But, they add, a steady flow of less sensational and more

> constructive criticism of IBM has also surfaced on the network during the past year. Memos passed through the claim that [Data Processing Division] employees are increasingly working without adequate tools or computing power, and with little or no merit incentives or career prospects. . . .
>
> Several months ago, Wheeler (an IBM systems programmer) decided to package together some of the VNET "gripe mail" into a collection. . . .
>
> According to sources, Wheeler then removed the names from the memos and sent a copy of the package to each of IBM's top executives. [Emmett, 1981: 48, 54]

This example illustrates one way in which systems can affect organizations by changing channels of communication.

Under what conditions are newer communication technologies likely to facilitate changes in structure, power or culture? Some studies have found only those impacts related to communication efficiency improvements. For example, Lippitt, Miller and Halamaj (1980) studied an electronic message system used solely within a single department of a larger organization and found that it was used primarily for downward hierarchical communication. Leduc (1979) studied an electronic message system used by members of a single organizational unit in Bell Canada to communicate internally and externally with consultants and also found that the system increased superior/subordinate communication over time. (Neither study reported on communication prior to the installation of the system or on communication via other media afterward.) The results of these studies suggest that new communication systems used inside a single organizational unit may slightly reinforce, but not change, existing patterns of authority and power. By implication, then, system-induced changes in authority and power are more likely to occur when the system is available for use across departmental lines, as VNET was.

INTERORGANIZATIONAL SYSTEMS

Interorganizational systems link an organization with one or more other firms. In addition to interacting with organizational features inside any or all of the interconnected firms, these systems interact with the relationships between the firm that supplies the system and the firms that use them and those between these firms and their competitors. These interactions result in impacts on interorganizational dependence and basis of competition. The impacts of interorganizational systems are summarized in figure 3.6 and are discussed more now.

FIGURE 3.6
Interorganizational System Impacts

Related Operational Features	Impact	Nature of Impact
Relations with customers and suppliers	Interorganizational dependence	Greater independence versus tighter linkages with other firms
Relations with competitors	Basis of competition in the industry	Worsened versus improved competitive position

Interorganizational Dependence and Basis of Competition

Independence is a synonym for power; firms that are dependent on others lack power in their interorganizational relationships. Interorganizational systems have the potential to alter power and dependence relationships among the firms that build them and the firms that use them. Consider American Hospital Supply's computerized order entry and inventory control systems for hospitals.

> Historically, hospitals did business with everybody as a way to maintain competition, says American's Waller. But now the movement is the other way, and they are looking to concentrate their ordering and maximize their purchasing strength with fewer suppliers. ["Systems that Slash Costs," 1980: 76E]

The suppliers gaining business are those who provide interorganizational systems that link themselves tightly to the hospital.

> The payoff for American has gone beyond the revenues received from hospitals for using these systems. For one thing, hospitals that are linked to the company's computer are more likely to buy supplies from American. And the average hospital order over the ASAP system averages 5.8 items, which compares with an industry average of just 1.7 items per order and with an average of 2.4 items on a conventional order received by American. Revenues show the same trend, with ASAP customers spending as much as three times more annually than they did before using American's manual system. ["Systems that Slash Costs," 1980: 76A–76E]

That this type of system can alter the basis of competition in an industry is also clear:

> American's competitors are moving in the same direction. Such hospital supply houses as General Medical Corp. and the Ross Div. of G. D. Searle & Co. are also offering computerized order-entry systems that bypass written purchase orders and the U.S. Postal Service. And both companies are expected in the next year or so to market a materials management package specifically designed for hospitals. ["Systems that Slash Costs," 1980: 76E]

In summary, interorganizational systems may both tighten the relationships among buyers and sellers, and many give the linked firms an edge in their dealings with other buyers and sellers; that is, interorganization systems may have impacts upon the efficiency and competitive position of marketing channels.

A marketing channel is a set of interdependent organizations involved in the process of delivering a product or service to a customer. Some believe that competition occurs between marketing channels rather than between individual firms at the same level in the marketing chain—i.e., between two retailers or two manufacturers. The distribution of personal computers provides an illustration. Many manufacturers are currently engaged in building them, including firms such as Apple Computers, Tandy and IBM. However, distribution of personal computers also involves firms such as captive retail outlets, direct mail distributors, independent retail chain stores and authorized dealerships. It makes more sense to discuss competition in personal computers in terms of IBM plus its distribution channels and Tandy plus its Radio Shack stores than solely in terms of the IBM Personal Computer versus the Tandy TRS-80.

The various firms in a marketing channel work out a division of labor among themselves, but this division is vulnerable to events like the decision of a firm to pursue forward or backward integration. Systems also may alter the division of labor in an industry by changing the economics of performing certain activities at various stages in the marketing channel. An automated order entry system like that of American Hospital Supply may enable that firm to reap the advantages of a higher sales price without a prohibitively higher cost of sales by inducing the hospital to perform a large part of the labor that would otherwise have been done by American's salespeople.

The process here is one of reducing redundant activities between pairs of firms. In traditional transactions involving two firms, some time or activity is required of both parties: Salespersons must call on the customers who wish to place orders, or customers must call up (or write to) order entry clerks at the vending firm. With an automated system, the time and activity of one member of this pair can be dramatically reduced, often that of the firm supplying the system (and products) to its customers. This may reduce the size of the sales contingent in the original firm or eliminate the need for other firms to act as intermediaries in the marketing channel. One might expect, then, that the widespread use of such systems will gradually shorten or alter marketing channels throughout the economy.

It is important to realize that this process of reducing redundancy of activities between firms is quite different than reducing redundancy of activities within firms. Inside an organization, performing some activities more than once can be a sign of administrative inefficiency or waste. However, even if a firm has no (or little) internal inefficiency, it may still be redoing activities already performed by customers. Thus, when an order entry clerk receives a letter from a customer requesting products, this person must transcribe the information in the letter to get it into the company's system. Whether this entails writing the data on a special form or entering the data via terminals, it still means doing again something the customer's personnel have already done once. Attacking this area of redundant activity can mean new cost savings for at least one of the two firms, but even more, it may mean a change in the marketing channel.

CONCLUSION   READ THIS .

This chapter has illustrated the types of impacts that are likely to result when a system with certain design features interacts with its organizational setting. These areas of impact are summarized in figure 3.7.

The framework presented in figure 3.7 can be used in two ways: (1) to identify and document the impacts that have occurred in a case of system use or (2) to predict the likely future impacts of introducing a system into a specific organizational context. Each way of using the framework requires some detailed knowledge about the system design features and the features of the setting in which it is used or into which it will be introduced. The procedure for assessing or predicting the impacts of an installed or projected system entails three steps:

1. Classifying the system design features,
2. Exploring the related organizational features,
3. Identifying the area and direction of system impact.

The first step of a system impact assessment is to understand the functions that the system will serve from an organizational perspective. Does the system link more than one organization? If so, it is an interorganizational

FIGURE 3.7
Chapter Summary

System Type	Related Organizational Features	Impact
Operational	Work force composition Job design Organizational structure, work flow coordination Organizational culture	Job opportunities and career prospects Job content and job satisfaction Horizontal structure Social interaction patterns
Monitoring and control	Job design Organizational culture	Autonomy and control Organizational psychology Organizational performance
Planning and decision	Work force composition Job design Organizational structure, centralization versus decentralization Organizational culture	Job opportunities and career prospects Job content and job satisfaction Decision making Poor structure Power structure Politics
Communication	Spatial and temporal factors Communication channels and networks	Location of work Geographic dispersion Who communicates to whom
Interorganizational	Relations with customers and suppliers Relations with competitors	Interorganizational dependence Basis of competition in industry

system, affecting relations with customers, suppliers and/or competitors. Does the system follow the cybernetic model of standards, measures, comparison, feedback and reward? If so, it is a monitoring and control system. Does the system have several subsystems with different features? Does the system have most, but not all, of the design features of the type it most closely represents—e.g., a monitoring and control system with no specific goal or standard? These are the kinds of questions one asks to classify the system.

The second step is to explore the organizational features related to the system's design features. This entails describing the current (and sometimes also the historic) state of each related organizational feature listed in the appropriate row of figure 3.7. For a communication system, for example, the assessor of impacts must first describe the spatial and temporal dimensions of communication in the organization before the system is/was introduced. How many locations are there, how far away? How frequently do people communicate about what issues, and how long does it take for completion of the communication? What media are used and what problems arise? Communication channels and networks must then be described in a similar way. For a planning and decision system, the assessor must describe, for the parts of the organization affected by the system, the composition of the work force, the design of jobs, the structure and the culture of the organization.

The most important part of this second step is to get enough detail to understand the point of view of every group affected by the system. The challenge of this step is to limit the amount of detail so that the analysis is efficient and timely. Experience is the best teacher in this process. Guidelines for bounding the organizational context are analyzed and presented in chapter 8; but the examples described throughout can provide useful clues about the appropriate level of detail.

The last step in an impact assessment is to use the classification of system design features and the exploration of related organizational features to identify or predict the area and direction of system impact. The figures throughout this chapter may be used as a starting place, but knowledge of the specific setting must temper the final assessment.

A useful procedure for checking the completeness of an impact assessment is to review the conceptual levels at which an impact analysis can take place. Three conceptual levels of analysis fall within the scope of this book: (1) the individual, (2) the intraorganizational and (3) the interorganizational. Other levels are possible but irrelevant to this perspective—e.g., the technical and societal levels. To check an impact analysis, the assessor can question if the analysis has explored the effects of the system in question on the various individuals who use the system, on each of the departments or subunits affected by the system and on the organization as a whole in its relations with its external environment. A given system may not have impacts at all three of these levels; but if a system does have impacts at more than one level of analysis, the impacts may differ in nature and direction from level to level. A system with problems at one level may have beneficial features at another. For example, American Hospital Supply's ASAP system may increase the market power of that company vis-à-vis its competitors by creating stronger links with its customers. At the same time, it may change the balance of power among departments in the company. Finally, at the individual level of analysis, it may change the job design of the salespeople who call on the hospital customers.

There is no easy way to establish the priority of impacts at different levels of analysis. It is not possible to assume that features at the interorganizational level will automatically outweigh bugs at the individual level or vice versa. Familiarity with the examples in this book and experience in conducting impact assessments will facilitate both the desirable shortcuts in impact analysis and the inevitable trade-offs in system design.

How Impacts Happen

INTRODUCTION

Chapter 3 described the kinds of impacts that may occur when the design features of a system interact with specific organizational features. The obvious next questions are: How are potentially negative impacts prevented? How are positive outcomes ensured?

The basics of what to do about impacts are contained in the notion of interaction. Impacts occur when system design features interact with specific organizational features. When system features match organizational features, the interaction of the two is smooth; the resulting effects tend to intensify slightly existing trends, whether negative or positive, rather than to create a major shift in the direction of performance. However, when the design features of a system conflict sharply with those of its context, the interaction of system and organization is rough and painful. If the system survives with its design features intact, the resulting effects are likely to be dramatic and the nature or direction of organizational performance to be noticeably altered.

For example, when a system with decentralizing design features is introduced into a decentralized context, the adjustment of organization to system will usually be smooth, and the impacts of the system are likely to be a slight worsening or improvement of the performance of the organization. Whether the change is for the better or worse will depend on factors such as how much improvement in performance was possible and whether the design features were well enough conceptualized and executed to achieve an improvement. The same decentralizing system introduced into a strongly centralized organization will face an abrupt and difficult transition. The conflict between system and organization will be so intense that something will have to give: Either the system will be altered, covertly or overtly, to reduce its dissonance with the organization, or the organization will begin to change its form and procedures to resemble the system more closely.

Unfortunately for system builders, the system, rather than the organization, usually adapts when the two are at odds. Organizations that are managing to survive in the environment that usually faces them typically achieve a balance or harmony among their major components: structure, culture, people, other assets, systems and leadership (Kotter, 1978). The complementarity among the components reinforces each of them, making change in any one component difficult without corresponding changes in all the others. Thus, a system that deviates sharply from existing structure, culture and so forth is unlikely to survive as designed, unless corresponding changes are made, more or less simultaneously, in structure, culture and so forth.

At first glance, the implications of this interaction perspective for what to do about system impacts appear unsatisfying. To avoid significant detrimental effects, avoid building systems with features that are sharply out of line with existing organizational features. This approach is rarely acceptable to those who perceive a problem with current organizational functioning and believe that systems may help to provide a solution. For those who would build systems, the interaction perspective offers the advice that the system is only a small component of the change effort required to achieve the impacts that the designer desires.

What else is required? Answering this question not only entails knowledge about system design features, related organizational features and impacts but also requires an understanding of the process of translating designers' intentions into design features and installing the system in its context of use. This chapter explores the process of how impacts happen.

THE INTERACTION PERSPECTIVE ON SYSTEM IMPACTS

Impacts happen over the course of the system life cycle. During the life cycle, the idea for a system is developed into a package of theoretical design features that are then executed, more or less completely or accurately, in procedures and/or devices like computers and their software. At some point, the conceptual or realized system is introduced to the organization through requests for design assistance or through usage training, and the system enters the phase of its existence where it is used in an organizational context.

At each phase in its life cycle, the evolving system may encounter circumstances that influence the outcomes, whether these are organizational impacts from using the system, abandonment of the system or basic modification of the system's design features. The old professor joke (thinks A, writes B, says C, means D, student hears E, right answer is F) illustrates a few of the ways in which an idea may be transformed en route to action. Consequently, each stage of the life cycle has been singled out, examined for the pitfalls that may result in undesirable outcomes and subjected to prescriptions for good practice.

For example, with respect to the phase of introducing the system to the organization, the implementor is frequently cautioned about the pitfalls of user resistance and lack of senior sponsorship and is urged to avoid these by involving users in the design process and by seeking top management support (Lucas, 1982). With respect to the phase of translating design concepts into a realized system, the designer is warned against sloppy conceptualization of users' needs and poor human-machine interfaces and is exhorted to follow a particular method of systems analysis and to attend to human factors engineering (Shneiderman, 1979). With respect to the use phase, the key problem is the ill-informed user and the solution is education and training (Keen and Bronsema, 1982).

The pitfalls and prescriptions that emerge from examining an isolated stage in the life cycle can provide useful insight and guidance to people who would build systems. However, there is a basic limitation of the pitfalls and prescriptions approach: It does not pinpoint the origins of a problem to a particular life cycle phase. This may lead to solving the wrong problem by implementing the wrong prescription. How does one know, for example, whether user resistance was caused by:

· A bad design feature due to a failure to obtain user input,
· Poor execution of a good design feature due to improperly trained system developers,
· Users irritated by the failure to solicit their input about the system in spite of its good design features and good technical execution?

In contrast, the interaction perspective approaches the system life cycle in an integrated manner. Problems that appear when the system is used frequently begin when the idea for the system first originates. Poor system execution and installation may introduce new problems, but these compound, rather than merely add to, problems with a system design concept. Furthermore, in many cases, the problems that arise in all stages of the system life cycle are interrelated in fundamental ways. The same circumstances that give rise to the problems observed during system use are those that led to the design concept for the system and to a selected approach to designing and implementing a system. Consequently, knowing what to do about system impacts requires understanding where the idea for a system comes from and how it is built and installed. The answers to these questions can frequently be found in the relationships among users and designers.

The relationships enacted among users and designers in the process of building and using systems are varied and complex. This is because both groups, users and designers, frequently consist of multiple subgroups with substantial differences on almost every dimension of relevance to organizational analysis—e.g., personality traits, motivation, occupational category, or organizational location. Frequently, the true designers of systems—the persons who specify design features—are not methods and procedures specialists or systems analysts but individuals who may be designing the system for their own use and for use by other groups.

The remainder of this book is devoted to examining the relationships among users and designers and the ways in which these relationships affect what happens to systems and to the organizations that use them. Later chapters go into greater detail about the system-building and system implementation phases of the life cycle. This chapter focuses more heavily on the interconnectedness between the design intentions phase and the use phase— specifically, on the ways in which users' reactions to systems are related to designers' motivations and intentions.

The Use Phase

How users use systems seems to be the bane of systems designers' existence. However they use systems, it is never quite the way the designers had intended. They sometimes use systems partially, not as fully as desired. They sometimes use systems for purposes for which they were not designed. This occasionally includes abusive use, fraud or crime, as described in chapter 3. Users sometimes actively resist systems, performing acts intended to sabotage the system and have it removed.

Partial Use. Users have been observed, on occasion, to use systems partially. This behavior may consist of requesting or using only some reports and not others. It may consist of using a system with multiple subsystems only as an operational system, ignoring features intended to provide decision support. It may entail observing the letter of the system but not the spirit.

One particular example of the latter case comes to mind—namely, the use of parallel systems. *Parallel systems* usually refers to the process of running an old system simultaneously with a new one for a period of time until the new one is operating correctly. However, in several cases, the notion of parallel systems has taken on an entirely different meaning.

Consider the 3PA system developed at the JHM Corporation (Markus, 1979; 1981). One of two plants slated to use this production planning and profit analysis system spent several months trying to explain to divisional headquarters why they should not use it. It did not have certain features they needed; further, they already had a system that did what 3PA did and more. However, the division needed 3PA to forecast the profit for both plants together, so the Athens plant was politely requested to use the new system.

The Athens plant did begin entering production plans and production data into 3PA. After a while, however, the numbers began to look funny to people at divisional headquarters. They phoned Athens with questions. Athens responded by fixing up the data, but the sequence of funny-looking data, phone calls and fix up occurred several times. Finally, headquarters sent someone to Athens to find out what the problem was. It turned out that the people at Athens were not using the 3PA system in quite the way the designers had in mind because they were also using their old WIP (work in process) system.

The WIP was a weekly batch system. Changes in inventory positions were recorded on printouts carried around by production controllers, then entered once a week into the system. The new 3PA system required a daily update via interactive terminals. What Athens did was to feed data into both systems simultaneously on the time schedule of the old WIP. Therefore, data in the 3PA system were always about half a week out of date. (So were the data in the WIP system, but all the important changes were remembered or marked down on the paper printouts.) Because Athens' staff were not using the 3PA (for their own purposes), they took little care to ensure that changes were accurately entered into 3PA. This is clearly a case of partial system use that does not quite conform to what the designers envisioned.

Augmented Use. Sometimes users have used systems for more purposes than the designers had planned. This additional use may appear to produce both windfall benefits for the organization, but it may occasionally stretch both the system's capabilities and the designers' tolerance.

Gasser (1983) described an engineering system designed to produce calculations concerning the flow of liquid through pipes. The chemical engineers discovered that the system, while accurate for pipes carrying cold liquids, produced erroneous results for pipes with hot liquids. Nevertheless, the engineers discovered a way to use the computer program for hot pipes: They indicated to the program that the liquid was cold and made a manual adjustment to the result. They described their unorthodox use of the system as running the hot pipes cold. Gasser gives many other examples of the ways in which users adapt, augment or work around systems to make them do what they want.

Another situation illustrates an attempt by users, not necessarily successful, to twist a system until it met their needs rather than the needs of the designers. This example concerns the second plant in the JHM division that produced 3PA (Markus, 1979; 1981).

When people at the Capital City plant learned that their division headquarters planned to install the 3PA system, they responded quite differently than Athens. Unlike Athens, Capital City had no systems for internal use, and they saw 3PA as a way to get the systems they wanted. When the division manager asked for volunteers to help design 3PA, several people from Capital City came forward and took very active roles in the project. One of these people became the official 3PA project leader, until he was suddenly removed from office two years later.

In early 1974, the project team was formed and given the charter to develop a production control system "which will be compatible with the needs of all personnel in the division." These needs differed. Athens believed they needed nothing. Headquarters needed costing and forecasting. According to one of the project team members, Capital City needed:

> A womb-to-womb MRP [material requirements planning] system. Something which will take the production plan and a bill of materials and tell me when I've got to make it, and when I've got to ship it, how much to keep in inventory and when to order raw materials.

At a design review meeting in mid-1975, the project team (heavily represented by Capital City) proposed the development of an MRP system and estimated a two-year completion time. The division manager rejected the proposal and commissioned a modification of Athens' WIP that could be completed in one year.

At the next design review meeting in mid-1976, it became obvious to the division manager that the team was emphasizing improvements to the operational systems they would use and was neglecting the forecasting and costing elements needed at the division level:

> They fed back to us what they were doing, telling us what they wanted to do next. I said, "where are my needs? I want a management exception report for use by me and the plant managers. I want a tool to help me manage the division better." If we had listened to the system that Capital City proposed, we'd not have been able to do 3PA. They couldn't get together on it. [Division manager]

The division manager replaced the Capital City project leader with a man from his own staff. 3PA was built. Capital City did not get its MRP, but not for lack of trying. The plant manager later told me that Capital City would get the systems they wanted "if it takes us twenty years."

This example illustrates how users often take systems (even in the design phase before they became operational) and try to turn them to fit objectives different than those of the original designers. This process often results in modifications to the original system, which over time can transform the system into something quite different than what it originally was.

Resistance. Designers have a tendency to apply the term *resistance* quite freely to any unintended (by them) use of systems. Often, what they label resistance a more generous observer would call partial or augmented use. In many cases, the label has been erroneously applied to innocent users who lack training on the system or who lack time away from daily job pressures to learn how to use it. The difference between these uses and resistance is that the latter term implies both actions and intentions aimed at sabotaging the system. A detailed example of resistance is given later in the chapter in the case of the Financial Information System (FIS).

The Intention Phase

Designers are frequently quick to attribute partial use, augmented use, abusive use or resistance to characteristics of the users. Users hate change. Users have cognitive styles that predispose them not to use systems or to use them ineffectively (Keen and Bronsema, 1981). Users lack experience with computers and may even fear them. [Don't laugh! It even has a name now: "cyberphobia," fear of computers. Someone has estimated that 5 percent of the nation's office workers "show traditional symptoms of severe phobia: nausea, hysteria, vertigo, stomachaches and cold sweats" when they have to deal with a computer at work (James, June 8, 1982).]

However, these same designers fail to see the seeds of user behavior in their own designs and intentions. Designers have been observed to build systems that were never really intended to be used. They have been documented as designing systems as weapons in the conflict between workers and managers. And they have been found occasionally to design systems to increase their power and control over other groups of users.

Symbolic Systems. Sometimes the point of a system is not to rationalize work, to reward or motivate people, to aid thinking and decision making or to decrease the constraints of time and space but to appear to do these things without doing them. People, groups and organizations that are vulnerable to control by outsiders have occasionally developed systems that make it look as if they are doing such a good job of managing that external intervention or control is unnecessary. In some cases, the system or its results may actually be used, but in other cases, it may not be, either because the designers do not wish their behavior to be regulated by outsiders or because the system, appearances aside, is not really capable of achieving its apparent purpose.

This description of the symbolic purpose of systems may conjure up Machiavellian and anti-utopian images. It suggests the science fiction story, whose title is long forgotten, of a future society governed according to the printouts of a massive, all-knowing computer. Not until somewhat later in the story is the reader made privy to the fact that the programmers had not finished their work: Behind the glittering facade of blinking lights and colorful displays sat a beer-drinking fellow in his undershorts who banged out replies to inquiries on a portable typewriter (a delightful personification of the phrase "garbage in–garbage out"!).

However piquant these images may be, they are quite unnecessary to an understanding of the pressures that produce symbolic systems. Consider a welfare agency that depends entirely upon the generosity of outsiders for the funds it needs to go on doing business. The heyday of enthusiasm for social programs has passed. Available funds from government sources have diminished, increasing the competition for private funds. People who disburse these funds find themselves in the uncomfortable position of trying to decide which cause is more worthy (multiple sclerosis research or alcoholism rehabilitation) and which agency is doing a better job of achieving its objectives (is the relevant criterion more people treated with fewer resources?). At the same time, they are increasingly sensitive to potential welfare frauds.

In this environment, an agency may well have difficulty attracting needed funds. The agency without a large endowment that cannot convince funders that it is doing a good job deploying its resources to achieve its goals does not have a chance of remaining in business. An automated system that tracks welfare cases and resource expenditures is certainly impressive evidence of the commitment to use resources wisely and well, as long as one does not inquire too closely about whether or not the system is being used.

Kling (1978b) studied such a system and found that agency personnel faithfully and enthusiastically supplied the input data and read the output reports but that the system did not actually affect internal operating procedures. He concluded that the system was believed to be so successful because it offered the agency a distinct advantage in competing for funds. It may never have been intended to improve internal administrative efficiency.

Another example of a symbolic system is the Program Evaluation and Review Technique (PERT) systems designed and used on the Polaris missile project:

> Construction of Polaris is an example of brilliant management, and one instance of this brilliance was to be known as an organization with brilliant management so that external agencies would leave it alone. When asked if they would use PERT, Polaris' managers said they would not use a formula for anything important. Rather they told someone to develop a method that would look scientific so that innovative management could be cited as a rationale for escaping outside control. [Wildavsky, 1978: 79]

A third example of symbolic systems can be found in this brief article from the *Wall Street Journal:*

> MEET OUR CHIEF DESIGNER, R2D2. Computers are taking over the work now being done by draftsmen. Many of the top architectural and engineering firms plan to buy computers that can graphically display building designs. The Paper Plane, a newsletter on automation, says, "By the end of this decade, more than 90% of these design firms will be using computers to do their work." Nevertheless, many architects aren't convinced that computer-aided design is worth the expense. Says an official with one firm, "We're getting one because our clients think any firm that's any good has one." [March 10, 1982]

Systems as Weapons in Class Conflict. Designers may also sometimes build systems to gain power in the struggle between workers and managers. According to one view of the world, the interests of management and labor are opposed to each other; management introduces technology like information systems to reduce dependence on, and to increase power over, labor. Consequently, according to this view, wherever two technologies exist to accomplish the same objective, the one adopted will have design characteristics that are more consistent with the objective of achieving or maintaining a dominant power position vis-à-vis labor.

Noble (1979) makes a compelling argument with respect to the case of automatically controlled machine tools. He documents the existence of two ways to program machine tools. One, the record playback method, records the motions of a skilled machinist as he or she makes a part. Once the first piece is made, subsequent parts can be made automatically by playing back the recorded magnetic tape. This, Noble argues, allows machinists on the shop floor to retain control over the new technology. The second method, numerical control, takes data from engineering drawings and converts them into a mathematical description of the part and of the machine actions needed to produce it. The programming is usually performed by engineers in an office away from the shop floor, which allows managers greater control over the technology and the machinists. Through historical analysis, Noble demonstrates that both methods were known and available during the early days of automated machine tools but that the numerical control method rapidly became dominant, largely because managers wished to release themselves from dependence upon skilled machinists.

> It is no wonder, then, that ... N.C. was often referred to as a management system, not as a technology for cutting metals. ... N.C. seemed to eliminate once and for all the problems of "pacing" [workers restricting output], to afford management greater control over production by replacing problematic time-study methods with "tape-time" (using the time it takes to run a part tape as the base for calculating rates) and replacing skilled machinists with more tractable "button-pushers", who would simply load and unload the automatic machinery. [1979: 337–338]

Similar examples of the use of computer technology as an instrument in class conflict are provided by Braverman (1974) and Kraft (1977).

Political Systems. Designers have been known to build systems to change the balance of power among departments or other units in the organization. Because this design intention is not viewed as legitimate in most organizations, systems designed to increase one party's power and control over others are frequently masked in the language and appearances of improved efficiency or decision-making performance. The example of FIS presented later in this chapter provides a dramatic example of a politically motivated system.

User-Designer Interactions

The interaction perspective looks for the origins of users' behavior with systems in the designers' intentions as embodied in system design features. These intentions are frequently a product of the relationships between users and designers and designers' ambitions to alter these relationships. The primary assumption of the interaction perspective is that an information system embodies a distribution of intraorganizational power among the key actors affected by its design. Intraorganizational power is an attribute of individuals or subgroups within the organization, like the head of marketing or the marketing department. Power can be defined as the ability to get one's way in the face of opposition or resistance to those desires (Pfeffer, 1981). An individual or subgroup acquires power in a number of ways, including personal characteristics like being an expert or being charismatic, but position in the formal structure of the organization often provides greater access to the resources that confer or symbolize power and the legitimacy required to use them.

Pfeffer (1981) describes the major structural determinants of power in organizations: dependence of others on the power holder, ability of the power holder to provide resources, ability of the power holder to cope with uncertainty, irreplaceability, and ability to affect a decision-making process. These determinants of power are relevant to the implementation and impact of systems, but the most frequently cited is the ability to cope with uncertainty. The reason for this is that the raison d'être of management information systems is to provide managers with useful information, allowing them to cope better with production problems and external events.

The information required to cope effectively with uncertainty is distributed through organizations in a nonrandom way; some people and groups have more access to it than others, and this gives them power. Some systems have design features that redistribute the information required to cope with uncertainty; thus, these systems have the potential to alter bases of power. Power redistributions due to systems have been documented by Kling (1978b), Bjorn-Andersen and Pederson (1980) and Hedberg et al. (1975), among others, and Bariff and Galbraith (1978) have reviewed the research and theoretical arguments about power shifts.

Thus, for example, a relatively stable balance of power will develop in the relationships between the purchasing, engineering, operations and production control departments in any manufacturing organization. Sometimes engineering will call the shots, sometimes manufacturing. The introduction of a new logistics system may funnel all key information through the production control department, thus giving them an unaccustomed power edge in their dealings with other groups. The result might be a permanent redistribution in the balance of intraorganizational power, unless something happens to prevent it. Resistance or unintended system use by those who stand to lose in the reallocation of power is the something that often prevents a power shift from taking place.

The interaction perspective allows some precise predictions about where resistance is likely to occur during the implementation of systems. Power, as it has been defined here, is a valuable resource. People and organizational subunits may differ in the extent to which they actively seek to gain power, but it is unlikely that they will voluntarily give it up. When the introduction of a system specifies a distribution of power that represents a loss to certain participants, they are likely to resist the system. Conversely, when the distribution of power implied in the design of a system represents a power gain to participants, they are likely to engage in behaviors that might signify acceptance of the system—namely, frequent use and/or positive statements about it. In general, neither would one expect people whose power is lessened by a system to accept it (gracefully) nor those who gain power to resist. In addition, the strength of resistance can be expected to vary in proportion to the size of the loss and its perceived importance.

Some specific conditions in the design of systems that produce losses or gains in power can be spelled out. Access to information is probably less important as a basis of power than the ability to control access to information or to define what information will be kept and manipulated in what ways (Pettigrew, 1972; Pfeffer, 1978; Laudon, 1974; Kling, 1978b). When a system centralizes control over data, the individual or subunit who gains the control is likely to accept the system readily, while those units losing control are likely to resist, even if they receive access to greater quantities of data in return. Similarly, decentralization of control over data is likely to be resisted by the unit formerly in control and to be accepted by units that gain control.

If control over data (whether centralized or local) has prevented certain groups from obtaining needed or desired access to it, the distribution of data, even unaccompanied by control over it, will provide the new recipients with the means of gaining power. Their prior dependence on the controlling group will be reduced since they will have an alternate source of data. Consequently, they are quite likely to accept a system that distributes access to data. In contrast, those whose monopoly over data is threatened are likely to resist. Distribution of data, which makes the performance of a subunit more visible and vulnerable to control attempts by other units, is likely to be resisted by the group whose performance is exposed (Lawler and Rhode, 1976) and accepted by those who would like to influence the others' performance.

The strength of resistance is also likely to be affected by the position of the person or subunit to whom one loses power. If the so-called winner is located in a vertically superior position in the hierarchy, resistance is much less likely than if the winner is a peer. Formal authority relationships tend to make power differences between superiors and subordinates more legitimate than similar differences among groups at the same horizontal level in the organization.

Resistance, as discussed here, is neither good nor bad in and of itself; whether or not it is so labeled usually depends on the vested interests of the person or group doing the labeling. Resistance can be an important, even a healthy, phenomenon by signaling that a system is altering the balance of power in ways that might cause major organizational dysfunctions.

Resistance is no more inevitable, no more an absolute characteristic of users or systems, than are positive or negative impacts. Resistance is an expression of choice about how to use a system, just as the design features represent the choices made by designers, as Noble says so well:

> To say that technologies embody human choices and that these choices reflect the intentions of the designers . . . is not to say that these intentions (desired impacts) are automatically fully realized in the simple construction and use of the technology. . . . In actuality, the "impacts" are always determined subsequent to the introduction of the technology and in ways not altogether consistent with the intentions of designers. . . . The relation between cause and effect is always mediated by a complex and often conflictive social process. . . . Often mistaken for (or dismissed as) "irrational fear of change" is the quite rational struggle on the part of those affected by the technology against the choices embodied in it and the impacts implied by them. . . . [1979: 320]

However, these choices by users and designers are not independent, not unconnected. They are both the product of an organizational context of structure and culture that describes and defines the relationships among users and designers.

THE IMPACTS OF FINANCIAL INFORMATION SYSTEM (FIS): THE CASE OF GOLDEN TRIANGLE CORPORATION

The case of FIS (Markus, 1979; 1983a) illustrates clearly the interconnections among problems appearing at each stage of a system's life cycle. It demonstrates unintended system use and shows the relationship of resistance to the system design features that embodied designers' intentions. It presents the context of relationships among users and designers that gave rise to both the design features and the resistance.

In order to make these points most sharply, the case of FIS is presented in two parts. The first part outlines an impact assessment of FIS, as described in chapter 3. The design features of FIS are described and related to the relevant features of the organization in which it was designed and used. The impacts of FIS are then discussed as the result of an interaction of system design features and organizational features.

The second part of the FIS case describes the entire life cycle of the system: designers' intentions, execution of these into a concrete system design, implementation or installation and use, including resistance and the subsequent modification of the system. This historical slant on FIS shows the value of the interaction perspective in explaining the outcomes and impacts of systems in terms of the relationships among users and designers.

FIS Design Features

The FIS system collects and summarizes financial data for the Golden Triangle Corporation (GTC). The system records transactions involving revenues and expenditures, assets and liabilities. It produces monthly profit and loss statements for each division and for the corporation as a whole; balance sheets are also produced by the system. The information collected by FIS is used primarily for external reporting and for managerial decision making.

FIS redesigned and replaced accounting operations and financial reporting procedures in GTC. A memo to GTC's top management, written before the system was built, described the benefits the system was intended to achieve:

1. Accelerated monthly closing.
2. Availability of detailed summary data.
3. Flexibility of reporting.
4. Flexibility of forecasting and budgeting and changes.
5. Great accuracy in data collection and reporting.
6. Data integration.
7. More effective utilization of computers.
8. Historical data availability.
9. Decreased clerical effort. [from internal GTC memo, 1972]

Because FIS rationalized and routinized the physical and some of the intellectual aspects of accounting work, the system can be classified as having the design features of both an operational system and a planning and decision system.

Related Organizational Features

The design features of FIS affect external organizations like the Securities and Exchange Commission, managers in every GTC subunit and the accountants in GTC who design accounting methods, gather accounting data, analyze and interpret financial conditions and communicate these to GTC's managers. For brevity, the impact assessment presented here is restricted to the accountants, who are the primary users of the system. Because there are two groups of accountants within GTC, differing in job design, structural relationships and culture, this restriction does not preclude an interesting analysis of the organizational features related to FIS.

FIS affects GTC's corporate accountants and divisional accountants. The key job tasks of corporate accountants are to develop guidelines for accounting practices for all subsidiaries and divisions within GTC and to consolidate their financial results. In contrast, the key job tasks of the divisional accountants are to collect financial data, to prepare financial statements, to analyze financial performance for their divisions and to report results to the corporate accountants for consolidation.

The two groups of accountants also differ in their structural positions in GTC. The corporate accountants report to the corporate controller, who reports to the vice-president of finance, who in turn reports to a GTC president in charge of staff units. In contrast, divisional accountants report to divisional general managers through divisional controllers (see figure 4.1). Corporate accountants, then, do not have line authority over the divisional accountants and, thus, cannot give them direct orders. However, the two groups are related laterally by a so-called dotted-line relationship, which conveys the advisory role of the corporate accountants.

The accounting groups also differ in culture. Corporate accountants describe themselves as financial accountants; divisional accountants describe themselves as managerial accountants. The distinction centers on the time frame and focus of their efforts. Corporate accountants focus on past events, summarized for outside groups; divisional accountants focus on current and future events, projected and evaluated for internal managers. At the time when FIS was installed, there was little or no job mobility between the two groups. Had corporate accountants been recruited from the divisional accounting ranks, the perceptual and cultural differences between the groups might not have been so great.

Impacts

FIS interacted with the organizational features of GTC in ways that produced changes in the accountants' jobs, in the structure of accounting operations and in the cultural patterns of interaction among the accountants. FIS changed the jobs of both groups of accountants: those of the financial accountants for the better, those of the divisional accountants for the worse.

FIS introduced the corporate accountants to the features of automation. They had formerly prepared the corporate consolidation by hand; FIS now performed this automatically. In addition, corporate accountants discovered that the system could perform tax accounting, a feature they had not anticipated. GTC was going through a period of rapid acquisition and divestiture at this time. Whenever a division or subsidiary was bought or sold, all of the corporate statements for the month, year and sometimes previous years had to be revised. Different types of units had different tax status (wholly owned domestic division versus international subsidiary), further complicat-

FIGURE 4.1
Golden Triangle Corporation Organization Chart, 1978

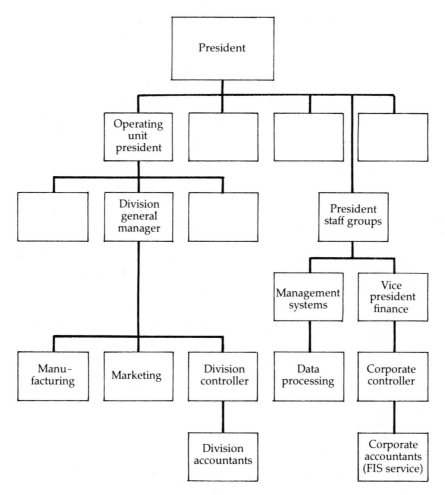

ing the process. FIS made light work of these job tasks of the corporate accountants.

In contrast, divisional accountants experienced FIS as a package of bugs when they tried to incorporate it into their daily routine. Prior to FIS, each division had had its own accounting system, many of which were manual. The most sophisticated was a computerized system that identified accounts with an eight-digit number. One measure of the difficulty of using FIS can be seen in the fact that it had twenty-four-digit account codes.

The process of entering data into FIS was very burdensome, a problem experienced only by divisional accountants who had the task of entering data into FIS. Corporate accountants interacted with the system only to retrieve reports, not to enter data. For example, before divisional accountants could post their daily transactions to accounts, the accounts had to have been defined to the computerized system in advance. Defining the accounts was not a one-time effort since new accounts were frequently set up for special purposes. Defining an account entailed relating it to all others in the chart of accounts and conveying this information to FIS through a special computer run, conducted only once a week. Creating the new accounts was not quite as difficult as one might suspect from the fact that twenty-four digits can specify an awfully large number of accounts. But the available documentation on FIS did not explain how to set up accounts. The divisions had to figure this out for themselves. One division even wrote its own training manual.

The problems did not end here, because the people entering data did not always know when new accounts were properly defined. If they entered data into an undefined account, the system kicked out the amount of the entry into a suspense account where it would remain until someone discovered the problem, set up the new account (several days later, since the account creation run was only once weekly) and then reentered the data. For a while during the early days of FIS use, the suspense acounts were larger than the regular accounts. Closing schedules were not relaxed to accommodate these difficulties; the divisions were expected to produce their statements as usual.

FIS, then, created problems for the divisions where they previously had had none. Further, the system did not help them in areas where they needed or wanted help. Collecting data for financial statements was only part of the divisional accountants' job. Primarily, they saw themselves as managerial accountants, attempting to provide to their divisional general managers timely forecasts about future costs and profits. As they saw it, the primary thrust of FIS was historical reports of past performance to outside agencies; in a phrase, FIS did financial accounting. To perform their managerial accounting, divisional accountants had a separate system (the Product Gross Profit system) that had interfaced with their old financial systems. FIS was supposed to work with the PGP system, but the divisions found themselves doing a large part of the translation by hand.

> FIS does not provide us with the data we need to prepare profit center reports. To prepare profit center reports, we must maintain a separate system, the PGP system. . . . They tell us we can use FIS for profit center reports! That's garbage! You *could* do it, but I've already told you how you have to enter data into FIS. To get a profit center report, you'd have to enter each transaction by commodity code. There are a thousand commodity codes. This would be a horrendous job. Besides, PGP is our product gross profit report. We've had this system unchanged for almost ten years.

... Naturally, the profit figures from this and FIS should reconcile, but they never do, so we have to make the necessary adjustments. [Divisional accountant]

In addition to its impacts on the jobs of the accountants who used it, FIS also affected the structural relationships among them. Prior to FIS, information in GTC was decentralized. Each division had a separate data base of financial information, controlled by divisional accountants. Corporate accountants and managers had access to summaries prepared by the accountants, but raw data were available to them only through special arrangements with divisional accountants. Any analysis capable of producing an interpretation different than those in the divisional accountants' summaries would require access to raw data. Thus, because they controlled access to the raw data, divisional accountants were located in a structural position of power with respect to the corporate accountants. They could control the interpretation drawn by the latter from financial data.

FIS disrupted this structural arrangement by creating a central data base of raw data under the control of corporate accountants (see figure 4.2). While divisional accountants could, of course, gain access to their own raw

FIGURE 4.2
FIS Final Design

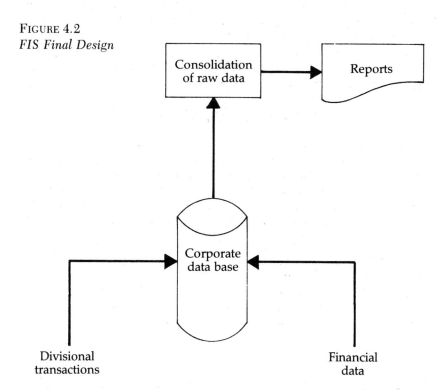

data, so could corporate accountants. This effectively eliminated one base of power the divisions formerly had in their dealings with corporate.

A change in formal organizational structure followed hard on the heels of the installation of FIS in January 1975. A few months later, several staff accounting functions that had formerly reported directly to the divisions were reorganized to report to corporate accounting (compare figures 4.3 and 4.4). This restructuring effectively centralized staff accounting functions, just as FIS centralized accounting data.

Many people within GTC denied that FIS had been responsible for this change in structure. However, they agreed that the structural change served a function similar to that of FIS. One corporate accountant said, "If the reorganization had occurred several years previously, FIS might never have been instigated. The reorganization eliminated much of the need for FIS."

FIS also had an impact on the culture of GTC: It affected the patterned relationships between the two accounting groups. The cultural impacts of FIS manifested themselves in behaviors, by divisional accountants, that are typically labeled "resistance." FIS started up in January 1975 in the largest division of GTC. In October 1975, an accountant from this division wrote a

FIGURE 4.3
Golden Triangle Corporation Organization Chart, 1968

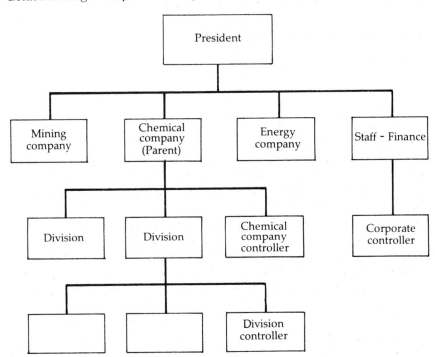

FIGURE 4.4
Golden Triangle Corporation Organization Chart, 1975

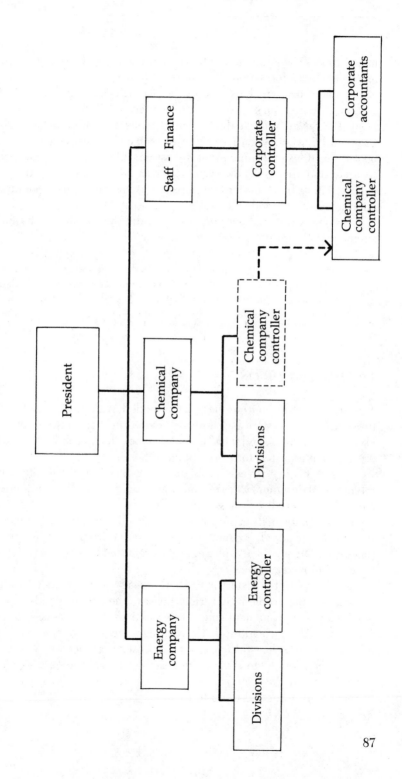

memo to the corporate accountants complaining that "except for providing more detailed information, the FIS system has not been beneficial to us."

Later divisional users were apparently no happier with the new system. One division kept on using its old accounting methods, after it started using FIS, even though this required twice the effort in recording data. Frequently, whenever there were discrepancies between the two sets of books, this division claimed that its system (thick manual ledger books) was accurate and that the new system was at fault. This recalcitrant division persisted in its behavior for two years until someone from corporate accounting carried the old ledgers away.

Some divisional accountants also admitted to slight data fudging to circumvent the problems with the system:

> If it turned out that an account we needed had not already been defined to FIS, rather than wait for the special account creation run, we might change the plant code, just to get the data in. After all, we knew what the numbers really were!

THE HISTORY OF FIS

The story of FIS neither began nor ended with the impacts described previously. The system had a history rooted in the relationships among its users and designers, and it had a sequel of modification and refinement as the organization adjusted to it over time.

The history reveals problems at every step of the life cycle. The intentions of designers were to gain control over users. These intentions were painfully executed into a system with poor human factors and technical difficulties on top of its organizational designs. During implementation, users were not consulted, trained or given relief from pressing operational schedules. During the use phase, users attempted to sabotage the system and have it removed.

However, the history of FIS also reveals the fundamental interconnections among the problems in various phases. The technical redesign of FIS to use data base management concepts was the rationale necessary to cover the execution of design features that gave designers control over users. The failure of designers to solicit design input from users was necessary to avoid pre-installation attempts to change the design features that were so offensive to the users. And the users' resistance is clearly related to what the design features were intended to do to them.

Intentions

GTC was formed in the late 1960s when the Golden Chemical Company merged with an energy company and a mining company (figure 4.3). At this time, several corporate staff groups were formed, including a financial group to oversee and integrate the activities of the merged companies. The new top management of the corporation staffed these functions with people from all three companies, but some in the parent chemical company believed that the outsiders received a disproportionately large number of the choice positions. For example, while the corporate controller was a former chemical company man, his boss, the vice-president of finance, came from the mining concern.

The new corporate controller had the charter of providing broad policy guidelines about accounting practices within GTC. He had only a dotted-line relationship to the controller in each of the three (later two) companies and a tiny staff to aid him in this endeavor. Whether this made matters worse or better is unclear, but the chemical company controller had originally been his boss and the two had been peers as assistant chemical company controllers prior to the merger.

Looking at his new job, the corporate controller saw the following situation. The energy company was well managed and had a single, central financial accounting system. The chemical company had seven different automated accounting systems and an uncounted number of different manual accounting systems in its many divisions. This made it difficult for them to provide information to him in response to his requests. Further, it is perhaps not much of an exaggeration to say that the chemical company had been dragged, kicking and screaming, into the merger: People in both the chemical company and the corporate office reported that the chemical company deliberately withheld data from corporate and that bad blood existed between the two groups.

In 1971, the corporate controller formed a task force of corporate employees headed by a former mining company employee to design a single, integrated financial accounting system for the corporation. While the system was expected to provide many benefits from rationalizing and routinizing accounting work, there were also political motives for the system. Speaking about FIS years later, a corporate accountant said:

> [The vice-president of finance] felt that the companies were doing things behind his back and that he needed a better way of ferreting out how the knaves were doing in the trenches. A large part of the reason for initiating FIS was to provide this information.

The head of data processing at the time (long since moved to a new company) remarked:

> FIS was definitely established for political reasons. [Corporate people] wanted to take over the whole world. But the Chemical Company was making all the money at the time, and they were not about to let anyone take over. Therein started the wars between the Chemical Company and Corporate.

His words make it quite clear that FIS was a weapon in that war.

Execution

The idea for FIS originated in the corporate accounting department around 1971. A task force was formed to evaluate the need for such a system and to estimate its costs and benefits. This task force was composed entirely of people from corporate accounting, some of whom had considerable data-processing experience. In 1972, after the necessary investigations and approvals, the task force arranged for the purchase of a financial accounting package from a software vendor (much to the chagrin of GTC's internal data-processing department, who would have preferred to build it themselves).

The purchased software package had design features that mirrored the way in which financial accounting was performed at GTC (compare figures 4.5 and 4.6). The package computerized formerly manual data bases, standardized inconsistent summarization procedures and automated corporate consolidation. Nevertheless, the FIS task force decided to modify the package, ostensibly to make use of modern data base management techniques. In the process of modification, however, which took over two and one-half years, the design team also replaced separate divisional data bases with a single corporate data base (see figure 4.2). This modification made it possible for the corporate accountants to gain control over divisional data.

Implementation

The FIS design team failed to discuss system design features with divisional accountants, who might have been expected to oppose the modifications of the purchased package. In fact, the task force members solicited no ideas about the design of FIS from divisional accountants until 1974, when it was time to set up the data base.

FIS started up in January 1975 in the largest division of GTC. In October 1975, an accountant from this division wrote a memo complaining about the system. In response to complaints from this and other users, a study

FIGURE 4.5
FIS Initial Design

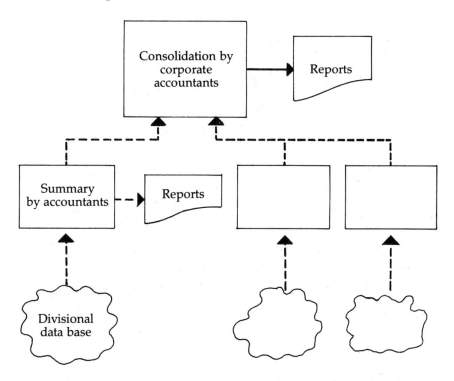

team was created to explore problems related to system inefficiency. The study team met for several months and made technical recommendations to the data-processing department. Execution of these changes proceeded slowly and received a setback in early 1977 when the data-processing project leader quit.

In the meantime, other divisions started up on the new system; all major divisions were using FIS by the end of 1975. This was surprising in light of the problems experienced by the initial users of FIS, especially since participation in the system was supposed to be voluntary. Many corporate accountants later pointed to this fact with pride as evidence of the success of FIS, but one person explained the incongruity as follows:

> Participation was voluntary on the surface, but there was a hidden inducement to participate. Those who wanted to wait to join FIS could do so, but they had to provide the same information manually. This would have been quite burdensome. So it really wasn't all that voluntary.

FIGURE 4.6
FIS Purchased Package Design

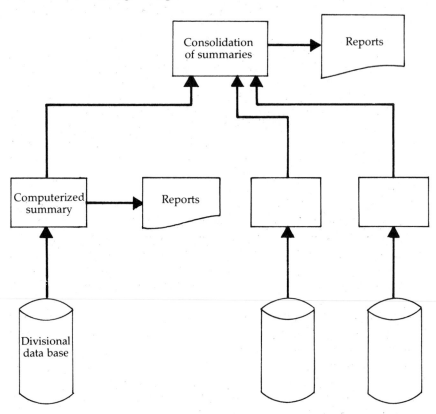

In August 1977, the memo-writing accountant in the first division to use FIS once again resorted to the pen:

> After being on FIS for several months, I expressed the opinion that the system was basically of little benefit. After two years and seven months, my opinion has not changed. Even worse, it seems to have become a system that is running people rather than people utilizing the system.

The corporate accountants found this behavior incomprehensible. One said, "I can't understand why the divisions don't like FIS. There are so many benefits." They bitterly denounced the trouble makers but agreed to a second task force to explore the problems with FIS. This one, which had representatives from divisional accounting, began meeting in December 1977 and met until March 1978 when it was disbanded by the vice-president of finance.

Use

An impact assessment, conducted well after the second task force was disbanded, made it clear that divisional resistance to FIS had not abated. This ongoing resistance was especially disturbing since the suggestions of the task force for improving FIS had been implemented. Divisional accountants still opposed the system, however. These facts argue strongly that the divisions were protesting not only the problems with the system but also the way in which the system altered cultural patterns and the distribution of power.

An analysis of interview notes, internal memos and task forces minutes, covering the period from 1975 to 1979, indicates that the difficulty of using FIS was only a secondary complaint; proposed changes in the way managerial accounting would be done was the real issue, one that no amount of technical fixing would solve. Further, this real issue was one of potential loss of power for divisional (managerial) accountants. Consider the following evidence.

First, an early memo about FIS (outlining a presentation to GTC's top management) explained the direction in which they were heading in the design of FIS. This direction represented a major shift in the way GTC did managerial accounting—that is, reporting to management about profit performance on specific products as opposed to the manipulation of aggregated, historical data. The intended shift in direction is clear in this excerpt from a 1972 memo:

> The last item of deficiencies that we list is the inability to analyze results on a total variance basis by business unit or corporate-wide. By that, we mean a lack of sales information by principal product and the lack of product line profitability. What was the volume of a given product? What was its price for a given period? What did that product contribute at the gross profit level? To me, the guts of our operation is what we do on a product line basis. In addition, we do not report on a given plant profitability. We feel that all this type of information, as was indicated, should all be part of a Financial Information System and available to management when needed.

Thus, corporate accountants had intended from the very beginning that FIS be used for managerial accounting, not just, as its name implies, for financial accounting.

Second, they did not immediately reveal these intentions to the divisions. When the staff in the divisions first heard about it, they were surprised. In the October 1975 memo complaining about FIS, the divisional writer noted, "I think we have to take a good look at what we have right now and improve it before we take any additional tasks proposed for the FIS system." The "additional tasks proposed" referred to product profit (managerial) accounting.

Third, corporate accountants were quite well aware that the divisions did not see eye to eye with them on the issue of managerial accounting. Recall that the second FIS task force was created in response to another angry memo written by the accountant in the first FIS-using division. Responding to that memo, a highly placed corporate accountant referred to the heart of the resistance issue:

> I must say that I am not surprised that your attitude toward the FIS system has not changed. . . . That same attitude is shared by the entire financial staff of your division, and hence, FIS will never be accepted nor will it be utilized fully as an analysis tool by your division. [August 1977]

"Analysis tool" here means a tool to be used in the analysis of managerial-oriented profit data. Note the use of the term *financial* to refer to the duties of divisional accountants.

Finally, the divisional accountants were quite explicit in distinguishing between operational and ease of use problems and use of the system for managerial accounting purposes. When the second task force was formed, it was partly "to improve things from a public relations point of view as well as from a technical point of view," according to one corporate accountant. But the divisional members of the committee did not intend to settle for symbolic gestures. "It was never really stated as such but one question we were looking at was: Should we look for a new system?" Task force minutes confirm this:

> During the sessions we have had thus far, one complex question already surfaced; is the system capable of being any more than a giant bookkeeping system, e.g., can it ever effectively serve divisional needs for budgeting, reporting, allocations, etc.? Therefore, we see two related issues we will attempt to offer recommendations on: (1) ways to deal with problems so the system can be counted on to operate effectively during month-end over the short-term, and (2) what, if anything, must be done to assure us that, for the long-term, we will have a system usable as more than a consolidator. [December 1977]

Since the task force was disbanded before it could tackle the second question, we will never know what it would have decided, but interview data suggest that the divisions remained very negative both toward FIS and toward the corporate accountants' proposed additional uses for it.

Here is the situation in summary. From the perspective of the divisional accountants, financial accounting is the legitimate domain of corporate accountants. A system intended primarily for financial accounting would have no real impact on the divisions, provided of course, that it was reasonably easy to use. The FIS system was not easy to use, but it was also not just a financial accounting system. It was intended to encroach upon what the divisional accountants perceived as their legitimate domain—that is, managerial accounting. Divisional accountants would resist the use of FIS for managerial accounting even if it were easy to use, and in fact, their resistance continued beyond March 1978 when the technical fixes were implemented.

Who won? Did the corporate accountants succeed in their attempt to alter the balance of power between themselves and the divisions? The answer is not altogether clear. The corporate accountants did succeed in removing certain accounting functions from the divisions and placing these under their control. And they managed to have the second task force disbanded in March 1978, before it could agree to disband FIS.

The divisional accountants, for their part, succeeded in redressing the more egregious faults of FIS and maybe in holding the line on corporate encroachments into managerial accounting territory. An administrator reporting to the president of one of GTC's operating groups summarized the feelings of many divisional accountants when he said, "I think it's about time they realized that FIS is really an operational tool. It just can't do everything."

In this remark, he summarized the view that FIS had been grudgingly accepted by divisional accountants as a tool for performing financial accounting (balance sheets, taxes and corporate consolidations) but that it was still being resisted as a managerial accounting tool. In contrast, the divisional accountants failed at having FIS removed. In all likelihood, the net result was something of a draw: The corporate accountants had better information than before, an important power advantage in their dealings with the divisions but not quite the total victory they had wished; the divisional accountants had regrouped and entrenched themselves to prevent any further losses.

When applied to FIS, the interaction perspective does more than identify and explain the resistance to and impacts of FIS. It also explains precisely why these specific design features of FIS were selected over others. Design features with the potential to cause a power shift in favor of the corporate accountants were chosen because the corporate accountants were motivated to increase their power and to control divisional accountants. Other systems, like the purchased package, would not go as far as the final design of FIS toward achieving these political objectives.

Further, the interaction perspective also helps explain the implementation strategies and tactics adopted by the corporate accountants to introduce FIS. The corporate accountants did not want suggestions from the divisions on the design of FIS because the divisions had different objectives and would probably have tried to select different design features. This point, stated differently, is that the same motivations that led the corporate accountants to select a set of design features that could create a power shift also led them to adopt an intervention strategy in which divisional accountants had little participation and influence.

CONCLUSION

The interaction perspective helps to identify and predict the impacts of and resistance to systems like FIS in their organizations. It explains impacts and resistance in terms of the relationships between users and designers. It shows the fundamental relationship between the intentions for systems and the strategies used to build and implement them.

In summary, the interaction perspective holds that impacts are changes in organizational features and functioning that result from an interaction of the design features of systems in their context. Design features are designers' intentions and desires for changed functioning, more or less accurately embodied in procedures and/or information technology. Resistance is behavior by users intended to avoid or alleviate the impacts of systems. Just as design features may result in impacts, resistance may result in changed system design features and, therefore, in different or neutralized impacts. Both system design features and resistance arise from the organizational context and history of relationships among users and designers. These factors also motivate the strategies for building and implementing systems, which in turn affect resistance and impact outcomes.

The interaction perspective has profound implications for how to avoid and prevent negative impacts. First, negative impacts can no more be prevented by good design features alone than positive impacts can be assured by good design features only. This is because no absolutely good design features exist—only design features that match or do not match the organizational features well. Consequently, while attention to human factors engineering and user friendly computer systems is obviously important (otherwise, no one will even bother to use the system and experience its potential impacts), this alone will not deter resistance or detrimental impacts.

Second, a good system-building process may result in systems that reflect more accurately the designers' intentions. However, this will not necessarily reduce resistance and negative impacts; it may even increase them. Resistance and impacts are products of what a system is intended to do, not how well it does it.

Third, a good process of system implementation will not necessarily result in achieving desired impacts or avoiding the undesired ones. Allowing users to participate in the design process may give them the opportunity to change or remove the features that would have achieved designers' objectives and to introduce new features that might achieve their own.

Fourth, resistance and negative impacts cannot be eliminated by actions that focus solely on the users: training, coercion or disciplinary procedures. The first simply does not get at the causes of resistance; the latter two drive it underground.

In summary, then, resistance and negative impacts cannot be effectively attacked by focusing independently on the problems that arise in each phase of the system's life cycle. Rather, an implementor of systems must create designs that take into account, in some way, the intentions, desires and motivations of all parties affected by the system; when the design differs from current organizational functioning, it must be reinforced by corresponding changes in other dimensions such as culture, structure or people. Furthermore, to assure that this design is accurately embodied in procedures and/or information technology, the implementor must also design a system-building process and an implementation process that is capable of producing and introducing the intended system. The next three chapters of this book apply the interaction perspective to the processes of designing and implementing systems. The final chapter addresses the design choices that face those who would build systems using the interaction perspective.

5

System Building

INTRODUCTION

System design features are the intentions of system designers more or less well captured and realized in procedures and information technology. But who are the system designers? Until now, the people designated as designers in this book have been those who make critical decisions about the design features of systems. These people are often managers and users of systems who are not computer professionals and who do not think of themselves as system designers. However, computer professionals also play an important role in the translation of intentions into the systems that are installed and used in organizations. This chapter and the next two are devoted to examining the role of these professionals and their contribution to the impacts and outcomes of systems.

The analytic techniques for predicting and explaining system impacts can also be used to predict and explain what happens during the execution phase, when ideas are converted into systems. The procedures and techniques for building systems are themselves systems with identifiable design features. System building occurs in an organizational context, peopled by vendors (chapter 6) and computer professionals (chapter 7) with differing interests and concerns. The design features of system building interact with the organizational context of vendors and professionals in ways that sometimes achieve neither the original design intentions nor the best efforts of computer professionals.

98

This chapter examines the design features of the system building process. Traditional system building is characterized by the complete specification of design features before execution, an emphasis on technical features rather than social, structural and political features and by professional control over user participation in design. However, other techniques for system building have been developed. Although not widely used, techniques such as evolutionary design, minimum critical specification and the ETHICS method achieve a greater fit with the interaction perspective than traditional methods. Nevertheless, they fail to address two dimensions central to the interaction perspective: fit or match with other organizational systems, like compensation, and structural or political relations between a department or a process and the rest of the organization.

TRADITIONAL SYSTEM BUILDING

The System Development Life Cycle

The system life cycle, discussed in chapter 4, consists of four phases: intention, execution, implementation and use. The execution phase, the subject of this and the next two chapters, has a life cycle of its own, often referred to as the system development life cycle. A breakdown of its phases includes analysis of requirements definition, design, implementation or code and test and maintenance.

Analysis produces a statement of what the system is supposed to do. To arrive at this statement, the analyst may perform a number of activities including documenting how the existing system (manual or computerized) works and determining users' needs through interviewing or other techniques (Davis, 1982). At a minimum, the analysis must address, at a general level of abstraction, the reports or other outputs desired from the system, the data or other inputs required to produce the outputs and the processing steps needed to transform inputs into outputs.

Design is the process of translating requirements into a plan for developing them as computer-based information or communication systems. This phase is believed to have two major subprocesses: preliminary (or logical) design and detailed (or physical) design. In logical design, the user's requirements for a system are translated into functional specifications: a statement of how the information system is supposed to do what it does. Lucas provides this list of the details required in a functional specification:

· The output (reports or CRT screens) from the system, its destination, uses, medium, frequency and examples;

- Inputs to the system, source of the data, medium of collection and/or input, specific items of data, estimated volumes and examples;
- Data storage files, medium, contents, record format, file structure, file size, estimated activity and update activity;
- Processing logic, flow, modules, processing mode, error conditions;
- Manual procedures around the system:

> The system design is not complete until the manual processing surrounding the computer system has been specified. What volume of activity will there be? How much time is required to perform the manual tasks? We should specify what processing the user has to perform and indicate the flow of information and the time required to complete the processing. . . . We should point out here that manual input procedures have caused many otherwise well-designed systems to fail. [Lucas, 1981: 214]

- Error control conditions and procedures, backup and security.

In detailed or physical design, the functional specifications are decomposed into a plan for programming. Various processing steps are grouped into blocks called modules, and the relationships between the modules are specified; in essence, physical design plans the sequence in which various procedures will be performed when software is being executed on a particular computer system. This step helps ensure efficient use of the computer and enables the programming task to be divided up among several people without many problems when the time comes to put it all back together. The output of this subphase is sometimes called "coding specs."

Implementation during the system development life cycle has a different meaning than implementation as discussed in chapter 4, where the term referred to introducing the system to the organization. During the implementation phase of system development, the software is coded or programmed, tested and installed—that is, made operational on a computer system. Preparing documentation of the system for ongoing maintenance is part of this phase. A key milestone in the implementation phase is acceptance testing, at which time the user decides whether or not the built system meets requirements and whether or not it should be declared operational. At this point, the system can be installed using any one of several strategies. For example, the new system may be run in parallel with the old until people are comfortable with its operation. The new system may be used on a pilot basis in one department until it is working satisfactorily. The old system may be phased out step by step as each phase of a new system is demonstrated to work. Or a go-for-broke conversion can be planned, replacing the old system with the new one overnight or over a weekend.

The life cycle of an application does not end with installation. The most obvious reason for this is the length of time that applications remain in use.

Gilchrest (1979) points out that "the lifetime of a major . . . system is typically on the order of 10 years." One rule of thumb is that an application can be expected to remain operational for one to two times the length of the development cycle; major applications frequently take two or three years to develop.

Once it is installed, an application system may be run on a computer system once a month, once a week, daily or continually. In the process, errors and exceptions inevitably arise. For example, if a transaction is improperly entered during keypunch or terminal entry, it must be ignored by the application program long enough for processing to be completed but must ultimately be corrected and reprocessed. Also, the process of testing rarely eliminates all software errors,* so these must be found and fixed when they make themselves known.

An important part of the ongoing maintenance of an application system consists of modifying it and enhancing it, not merely fixing errors. This may entail reformatting reports, adding new reports or changing processing to incorporate new legislation (e.g., payroll regulations) or responding to specific user requests. Modifications may include rewriting pieces of the software programs to make them operate more efficiently or to accommodate changes in hardware and systems software. Finally, major enhancements take place when the scope of an application is enlarged to include new functions, when previously separate applications are integrated and when applications are converted to run on a new computer system.

The Design Features of System Building

The system development life cycle may be viewed as an operational system intended to rationalize and routinize the process of building systems. The various techniques of system analysis may be viewed as planning and decision systems with the purpose of structuring the intellectual processes of determining user requirements. Similarly, various techniques for improving the coding and testing portion of the life cycle may also be classified according to the systems framework of chapter 2.

It is possible, then, to apply the procedure described in chapter 3 to assess the impacts of the system development life cycle. The first step, to

---

*Called bugs in the jargon after Grace Hopper discovered the cause of a short-circuit in an early computer system—a moth. If a program is even moderately complex, almost an infinite number of paths through it can exist. Not all of these paths can be tested. Once an application is in production, some program logic errors will be discovered through the occurrence of an unusual combination of data values. Some software errors are never discovered or, if discovered, fixed.

analyze the design features of the life cycle, follows. The second and third steps, to identify and describe the related organizational features and to assess the impacts from the interaction of system and organizational features, are presented in chapters 6 and 7.

One hallmark of the system development life cycle is its discrete stages. Before the design phase can begin, analysis must be complete. Before coding can begin, design must be complete. Passage of an incomplete system from stage to stage is typically worked by handoffs, signoffs or approvals of varying degrees of formality. It is usually also marked by the transfer of a system into the sphere of authority of another occupational specialty or organizational group, but this is the topic of chapter 7.

In practice, the life cycle stages of a system are rarely this discrete. The system may be segmented or decomposed into subsystems. Design work may proceed on the first system while analysis takes place on the second. In addition, errors or omissions in analysis may become apparent while the system is in design or coding; recycling through the earlier stages may be necessary to correct the errors. Finally, user needs may change while the system is under construction—for example, due to a change in governmental regulation—requiring iteration of the earlier developmental phases.

In spite of these daily adjustments, the life cycle imposes a definite discipline on the system-building process. It forces the system builder to specify completely (although this term is clearly subject to interpretation) what the system is supposed to do before construction begins.

There are strong practical arguments for this principle of full specification. Systems implemented in computer technology have proven themselves notoriously difficult to modify successfully; in many cases, the process of changing system functions, or even of fixing errors in operation, has introduced new bugs. Advances in computer and programming technology have been providing some relief on this score, but most computer professionals and information services managers are firmly convinced, from hard experience, that full specification of a system's design features improves the process of building systems. It shortens development time, prevents recycling and rework and increases the likelihood that the system built is the system intended.

Unfortunately, this belief is not fully supported by reality. A growing number of people are starting to question the utility of full specification, at least for certain classes of systems like decision support. These people argue that system designers and users cannot be expected to anticipate every use of the system, every function a system will be expected to provide. This is so in part because people lack perfect knowledge about future events. It is true also because of limitations on human conceptual skills. It is a common experience for people not to understand or not to be able to articulate what they are doing until they have done it. Even then, articulation of intuitive thought processes or of activities that have become habit or second nature may be difficult.

System designers have long recognized the pitfalls of the principle of full specification:

> For a manager to know what information he needs he must be aware of each type of decision he should make (as well he does) and he must have an adequate model of each. These conditions are seldom satisfied. [Berrrisford and Wetherbe, 1979: 12]

As the system analysts' adage would have it: "What users ask for is not what they want, what they want is not what they need" (Wetherbe, 1982).

A second design feature of the system development life cycle is its heavy emphasis on the technical aspects of systems. The jobs created or altered by new systems are rarely given the same attention as the systems themselves. Related organizational features such as structure and compensation practices are frequently not addressed at all.

When a system analyst using traditional design methods compares two alternative design concepts, the analyst evaluates the trade-offs between them in technical terms. Consider this example:

> In the design of a system to process accounts receivable on a minicomputer, a question arose as to how one could maintain an alphabetical sequence for a customer listing. Because the minicomputer had limited input/output capabilities and was slow in sorting, the analyst suggested that a numbering scheme be used to create the described sequence.
>
> By leaving a large number of digits between successive customers, the user could create a pseudoalphabetical listing. For example, if customer Adams were assigned number 1000, the next customer, Adamson, would be given 1010. If a new customer name Adamsen were added to the file, the number 1005 would be given. Over time, exact alphabetic order would not be maintained, but the results would be quite close. Certainly this process would be easier for the programmer and more efficient for the computer. [Lucas, 1981: 113]

In this example, the trade-off was ultimately decided in favor of the user rather than the computer:

> It was clear, when the analyst examined the cogent reasons of the user and empathized with her, that the tradeoffs should be decided in favor of sorting the customers alphabetically as the user requested. [Lucas 1981: 113]

Clearly, however, the choice is in the hands of the system analyst, whose predominantly technical orientation may tend to skew less clear-cut decisions toward the best technical solution.

The emphasis on the technical features of systems and the neglect of social and political organizational features is fostered and reinforced by powerful, practical concerns. The analysis and design of methods and procedures in organizations is often the responsibility of industrial engineers or organization and methods specialists. The design of organizational structures and compensation schemes is usually the responsibility of line managers or personnel and human resources departments. Professional system analysts or designers are not trained in these disciplines and would encroach on the territory of others even if they were.

Leaving these concerns aside, however, traditional methods of system analysis and design intend only to address technical system features. They ask no questions about organizational features and offer no suggestions for incorporating these in system design. They provide no guidelines for soliciting or including the input of specialists in methods or human relations.

Many traditional analysis and design methods do, however, provide guidelines to the professional designer for soliciting input from users (Lucas, 1981; Davis, 1982), but this feature of traditional system development has two problems. First, it assumes that users are the appropriate source of information about such factors of job design, organizational structure and compensation, an assumption which the case of FIS alone should be enough to eliminate. Users as system designers frequently have political interests and motivations that interfere with effective system design, and they are often as ignorant of organizational features as their counterparts, the computer professionals.

Second, traditional system development places the computer professional in control of user participation in system development. The professional selects the mode of participation (interview, questionnaire), determines who will participate, decides what questions to ask and interprets the answers. Combined with a bias toward technical concerns and a limited understanding of organizational issues, control over user participation in design contributes to a neglect of the organizational features related to systems.

In summary, then, the traditional system development life cycle has three critical design features. First, it requires full specification of system design features before the system is built. Second, it emphasizes technical features at the expense of social and political features. Third, it puts system design professionals in control of the nature and extent of user participation.

ALTERNATIVES TO FULL SPECIFICATION

A number of alternatives have been proposed to overcome the problems in the traditional development life cycle and in traditional analysis and user needs assessment. A complete review of the methods, their bugs and features, is neglected here in favor of a detailed description of a few of the more promising alternatives. None of the alternatives, however, is widely known or practiced, and none of them addresses the full range of organizational features emphasized in the interaction perspective. Two alternatives to the principle of full specification are evolutionary development and minimum critical specification.

Evolutionary Development

Variously described as heuristic design (Berrisford and Wetherbe, 1979), adaptive development (Keen and Gambino, 1980), evolutionary development, "middle-out" design and prototyping, the essence of this approach is that "it explicitly proceeds without [full] functional specification" (Keen, 1979). Consider this brief description of the process accidentally discovered in a major oil exploration firm:

> In an effort to demonstrate the capabilities of the new technology, the MIS group designed and developed a demonstration system to process a geophysical database . . . loaded with a sample of live records. . . . The demonstration system was functioning in less than a week. . . .
>
> The MIS group showed the geophysical personnel how readily modifications could be made to the system in order to accommodate additional requirements. . . .
>
> Geophysical personnel were impressed with the demonstration. In fact, they were so impressed that they learned how the sample data had been entered and began loading the remainder of the data.
>
> The MIS group quickly explained that the system was . . . not intended for production use. Geophysical personnel did not understand. The system was of immediate use to them as it currently functioned. [Berrisford and Wetherbe, 1979: 14]

As practiced explicitly, not accidentally, the idea behind evolutionary development is for the designer to present to users as quickly as possible a limited working system with which the user can experiment. Through hands-on practice, the users and designers can rapidly determine how close the demonstrated system or prototype comes to meeting users' needs. The necessary refinements can then be made incrementally. An additional feature of this iterative method is that it helps users to learn new procedures that they could never have specified in advance. The designer can capture these learnings and build them into the evolving system. In a very real sense, full functional specification of a system is the output of, rather than the input to, the system development process.

Minimum Critical Specification

Evolutionary development, however labeled, was proposed as a practical solution to the demonstrated inability of users to specify their desires and of designers to decide what the users need. Similar development strategies have been proposed by others for an entirely different reason, a more philosophical one but entirely practical in its implications. According to the principle of minimum critical specification, a good design specifies what is essential but nothing that is nonessential (Cherns, 1976). This achieves two benefits. First, it avoids closing off any options that can be left open so the people who perform the procedures have enough flexibility to respond effectively to unforeseen circumstances. Second, by building discretion, autonomy and self-control into people's jobs, this design principle enhances job satisfaction while it improves task performance.

The problem of overspecification and the solution of minimum critical specification can be seen in the example of a work design experiment conducted in Sweden in the late 1960s. The experiment took place in Department 698 of the Sickla Works of Altas Copco Mining and Construction Techniques, Inc. Its objective was to increase work satisfaction among the employees "although management had made it clear that the current level [of productivity] should at least be maintained" (Bjork, 1979).

> The department had recently been reorganized and rationalized on the basis of time and motion studies. Its personnel, picked for their experience, skill and diligence, had more than met the goals set for the rationalized department, which involved a 25 percent increase in productivity. . . .
>
> The department consisted of 12 men and a foreman. The input to the department consisted of the component parts of the rock drills, which came from elsewhere in the plant or were obtained outside. The department's products were some 40 different drill models, mostly variations on six basic types. . . .
>
> The department operated on a traditional "one man, one machine" basis. Each of the 12 workers had one of the following tasks: degreasing the incoming parts in a washing machine, grinding and honing the air throttle elements, preassembly, assembly, testing, painting and packing. . . . Most of the workers knew only their own regular job and perhaps one or two of the easier other tasks. [Bjork, 1979: 221-222]

This fully specified work organization was highly productive, but researchers believed that "the social system in Department 698 had become 'hung up' on the technological system in ways that contributed neither to fulfilling the unit's objectives nor to the workers' satisfaction" (Bjork, 1979).

> There was a great deal of concern and mutual suspicion with regard to rates of pay, and an informal status hierarchy had developed [among the workers] based on skill, job difficulty and pay levels. The two workers on the assembly line felt themselves most exploited: underpaid and bound to one location and to a flow of work they could not control. [1979: 222]

The researchers invited the workers to participate in redesigning their work setting:

> One major change was suggested by almost every worker [in a series of interviews]. It was to do away with the conveyor-belt assembly line and substitute a large table on which the drills could be assembled at the men's own pace and in their own way. The company agreed to try it, and after some delay a specially built table designed by the workers was installed and several lots of drills were assembled on it. Rather surprisingly the drills were assembled as quickly as they had been along the production line, even without a breaking-in period. The table made for freer and less monotonous work. On the other hand, it was hard to handle the heavy steel components and move them around and to keep track of the many small parts on the table. Slowly a new point of view developed among the workers. The conveyor belt was seen not simply as a source of unvarying stress but as a tool: it could be used as a means of transportation, with its speed controlled by the workers themselves as they assembled drills on benches alongside the belt in the new ways they had developed by working around the table. Elements of the belt were brought back, having taken on a new meaning for the workers as a result of their own active experimentation. [1979: 222]

Later, the workers of Department 698 began to experiment with pay systems and a division of labor consistent with a philosophy of specifying nothing more than the essential.

> The basic idea was that the workers in the department would constitute a single team, with equal pay. The team would be split up into groups of from two to four men, and each group would move through the department, carrying out the entire sequence of operations for a single lot, or batch, of drills. The allocation of individual tasks within each group would be left up to the group to decide, as would the exact method of performing each operation. It was to be more than job rotation among previously designed jobs, since the old designs could be scrapped in favor of whatever methods the men found most convenient and satisfying. [1979: 222]

The principle of minimum critical specification was also applied successfully in two units of the U.S. Navy, the Fleet Ocean Surveillance Information Facilities in Spain and Japan. These units were part of a network of analysis centers supporting the U.S. military command structure (Johnson and Burke, 1981). The network of surveillance facilities had been set up so that the same raw data would be sent to all units sharing an interest in the geographic region from which it originated. This redundancy helped to prevent oversight and to encourage multiple interpretations. The consequence was a spirit of competition among the facilities to report data and interpretations faster. Therefore, the facilities were quite receptive to new methods that improved the speed and quality of their analysis.

Each of the facilities consisted of a number of watch teams and a central reference section. The watch teams screened communications regarding maritime conditions in their environment and manually plotted onto charts those items they believed were significant. They drew inferences from positional data and published their evaluations in three kinds of reports: (1) daily scheduled summary reports, (2) messages about conditions requiring immediate attention and (3) responses to inquiries. The reference section, composed of the most experienced personnel in the facility, supported all watch teams by working on special projects and by providing continuity on certain issues. The teams were expected to stay beyond the end of their watch to complete all reports. Thus, watches much longer than eight hours were common. To assist in their tasks, the facility had access to three obsolete word processors, which produced punched paper tape input to the communication system. The equipment frequently broke down, sometimes erasing the work of others and requiring a reversion to an all-manual process.

The tasks of watch standing and analysis required considerable skill: It took at least three to five months of on-the-job training for a new person to become fully qualified. Decision makers in Washington, D.C., had become increasingly concerned about the decline in the number and quality of enlisted men and of officers electing additional tours of duty. Consequently, they decided to procure and install advanced analytical, text-editing and communications equipment to enhance the capabilities of the watch teams.

Initially, no one expected this replacement of equipment to effect a major change in the work of the watch teams, although a slight productivity increase was anticipated. However, the implementors elected an implementation strategy that did not mandate or specify the use of the equipment, in order to maximize learning and development of appropriate new methods. They did this by not replacing the old equipment with the new. The new equipment was installed side by side with the old, and trial use of it was mandated. All personnel were trained and encouraged to experiment with the equipment during a two-week shutdown of the facility for precisely that purpose. Even the software for the system was not fully specified in advance. A software professional was available during the shutdown to make modifications suggested by the watch teams.

The implementors intended the surveillance facilities to decide if they wanted to adopt the new equipment for exclusive use at the end of a six-month trial period. The old equipment was to remain available for use throughout that time. However, two days after the beginning of the shutdown period at the facility in Spain, the watch personnel dismantled the old word processors and plotting tables and began to use the new equipment exclusively. As they learned to use the equipment in their culture of friendly competition and experimentation, they discovered new ways to do their jobs that they shared with each other. For example, one team member:

> [F]igured out how to create some standard reporting formats and to store them. He then began to feed . . . data directly into the [word processor], store it in a [word processing] file, and compose his reports directly on the machine entering data from the file as needed. He began to complete and send his reports much faster than other people. Others adopted his innovation and added to it. . . .

> In addition to required daily reports, they had to respond to any number of special requests for information. Although several days might have been permitted . . . for an ad hoc inquiry, the management practice was that a request be answered as soon as possible, usually before the analysts left for the day. The ad hoc requests thus became prime targets for productivity improvements.

> Most personnel recognized that there was some redundancy in the kinds of questions they received. They kept folders of information gathered for past requests and frequently went rummaging for information to help them respond to new information. The notions of "file", "list", "format" which helped them reorganize their regular reports were applied to their ad hoc reports. They began to identify patterns of requests, to categorize kinds of information, and to organize their electronic files to automate the routine parts of ad hoc work. . . . A similar sequence of events happened in Japan. . . .

> Within six months it was uncommon for a person to take longer than eight hours to finish the work. More important than the reduction in the amount of time was the reshifting of how individuals spent their time at work. They were now able to devote more of their time to thinking about what was going on . . . because they spent less time physically transforming information from one shape to another. [And] they were able to experiment with the wording of their reports [shaping them] to their audiences' informational needs. [Johnson and Burke, 1981]

ALTERNATIVES TO TECHNICAL SYSTEM DESIGN

The technical emphasis of traditional systems analysis raises few concerns in cases where computer-based systems affect only a small portion of the jobs of a few people. However, when data collection and input or system usage fill one or more whole jobs, system implementors and designers should be concerned with job design and satisfaction. Mumford and her colleagues have developed an alternative design method that emphasizes considerations of work and job design (Mumford, 1971; Mumford and Weir, 1979). The method is called ETHICS—Effective Technical and Human Implementation of Computer Systems.

The ETHICS Method

The ETHICS method grew out of the authors' research on the implementation of computer systems.

> In some of the organisations we studied, the computer system was being installed in several locations, or a single computer system was serving several different locations in the same way, and yet the jobs of employees associated with these systems differed quite significantly in the different locations. Some of the jobs had much more interest and variety even though they were dealing with input or output from the same computer system. . . .
>
> In most cases, jobs were designed by the local user manager almost as an afterthought when the technical system was being implemented, and needed human assistance to complete the tasks which the computer could not do. But by that time all kinds of constraints had been unwittingly built into the technical computer system, and these considerably curtailed his freedom to organise the work in a way which would create interesting jobs. [Mumford and Weir, 1979: 1–2]

Accordingly, the ETHICS method was based on two assumptions:

> 1. That many kinds of technology (and this is particularly true of computer technology) are sufficiently flexible to allow for the design of systems which take into account the needs of employees for satisfying work. Therefore both the technical and human parts of a work system should be designed with this objective in view.
>
> 2. That even in situations where the technical system has been designed and implemented it is still possible to redesign jobs in ways which will make them more satisfying. [1979: 2]

Mumford and her colleague, then, advocate dual sets of objectives for any system implementation: first, technical and economic goals and, second, goals for improving the job satisfaction of those who work with and around the system. They define job satisfaction as a good fit or match between the expectations that employees bring to the job and the requirements of the job as defined by the organization. Employees' job expectations are partly a result of personality characteristics and partly conditioned by factors in the social setting of work and society at large. In all, they have identified five dimensions along which job satisfaction can be assessed:

- The knowledge fit—the degree to which the employee's job allows him to use and develop his skills and knowledge.
- The psychological fit—the degree to which the employee's job allows him to further his private interests for achievement, recognition, advancement, status or whatever.
- The efficiency fit—the degree to which the job offers financial rewards and supervisory controls which are acceptable to the employee.
- The task-structure fit—the degree to which the job meets the employee's requirements for variety, interest, feedback, task identity and autonomy in the job.
- The ethical fit—the degree to which the values and philosophy of the employer are compatible with those of the employee.

The essence of the ETHICS method is the identification of compatible pairs of alternative technical and social designs after establishing technical and social objectives. Mumford and colleagues suggest the following procedure for doing this. Users participate throughout. Technical and social alternatives are generated independently, perhaps simultaneously, by different groups of analysts. Then the technical alternatives are matched up with the social alternatives compatible with them, if any. This results in a set of feasible sociotechnical solutions, which are evaluated against the technical and social objectives for the system and ranked. A detailed design is then prepared for the best alternative, and if acceptable, it is implemented (see figure 5.1).

The process of specifying the technical and social alternatives is probably the most difficult part of the ETHICS methodology. Mumford and Weir (1979) allow that classically trained system analysts using traditional development methods can generate technical alternatives. They recommend, however, that the analysts go through a careful process of specifying technical needs, identifying technical constraints, identifying resources available for technical systems and specifying technical objectives before setting out alternative technical solutions. To identify social alternatives, a different procedure is used. The ETHICS method uses a questionnaire for assessing the level of job satisfaction in the organization prior to computer system design and implementation. Users analyze the questionnaire results to identify areas where job satisfaction can be improved and ways in which the design of the computer system might contribute to this objective.

FIGURE 5.1
The ETHICS Method

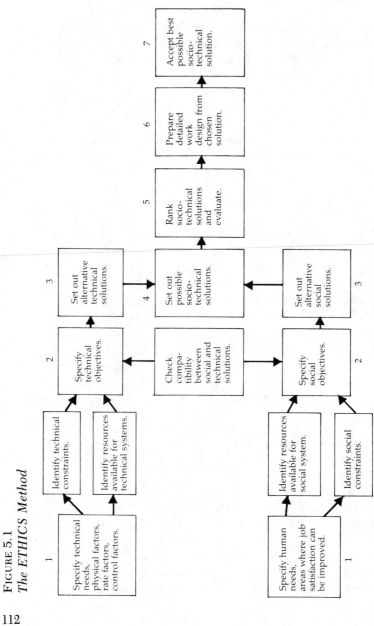

SOURCE: Enid Mumford and Mary Weir. *Computer Systems in Work Design—The ETHICS Method* (New York: Wiley, 1979: 37).

An Illustration of the ETHICS Method

Mumford and Weir (1979) have synthesized a case, summarized here, about the accounting department of a mail-order firm to illustrate their method. The technical objectives of this department were identified as:

· Reducing staff requirements by at least 200 people,
· Improving efficiency by increasing throughput,
· Reducing the number of errors,
· Processing orders fast enough for goods to be sent from the warehouses the day orders are received.

The technical constraints were listed as:

· Keeping the amount of breakdown time to a minimum, establishing adequate backup and recovery procedures,
· Choosing the least expensive equipment possible consistent with social objectives.

The available technical resources were catalogued as:

· Experienced systems analysts in the management services department, though more are needed,
· Assistance available from the computer vendor,
· Funds allocated by the board up to a budgeted limit.

Analyzing these needs, constraints and resources led to this statement of the technical objectives for the system: Reduce staff requirements, which are becoming increasingly difficult to meet because of a shrinkage in the labor market, and improve the level of service to customers.

Setting the social objectives for the system started with administering the job satisfaction questionnaire to people in the department. An analysis of questionnaire responses led to the conclusion that several social dimensions needed to be improved. Among these were:

Type of Fit	Unsatisfactory Aspects	Suggested Improvements
Knowledge fit	Skills and knowledge underutilised Work not interesting Work not difficult enough Not enough opportunities for self-development	Creation of larger jobs requiring greater use of different skills Job progression from simple to increasingly difficult

SOURCE: Adapted from Mumford (1979: 148).

The analysis also identified some strengths in the existing work organization that should be incorporated into the new system. Among these satisfactory aspects were the level of responsibility in the jobs, the level of pay (although the bonus system needed improvement) and feelings of achievement among the workers.

The social constraints on system design were identified as:

· Reducing staff by 200 without layoffs,
· Making jobs appropriate to skill levels of existing staff, bearing in mind their perceptions that skills are currently underutilized.

The resources available for the social system were identified as:

· Already available training facilities and the sanction to recruit two senior people for training,
· Available assistance from the organization and methods department,
· Funds for temporary staff during the changeover.

An analysis of these needs, constraints and resources led to the following formulation of human (job satisfaction) objectives for the system:

> The creation of a workforce which is efficient, motivated and identified with [the mail order firm's] interests and which has job security, job satisfaction and opportunities for personal-growth development. [1979: 155]

With both technical and social objectives specified, the next step in the method was to generate a list of technical alternatives that would meet these objectives within the constraints and resources available. At a minimum, the alternatives must specify methods for input, processing and output, where input refers to the equipment used to enter data, processing refers to batch or on-line computing and output refers to the medium and format of reports and other system products.

In Mumford and Weir's methodology, the type of computing equipment and technology used is important because of the implications of the equipment for job design. For example, key-to-tape input, in which an operator types information onto magnetic tape, has different advantages and disadvantages when considered from technical-economic or from job satisfaction perspectives. The technical-economic pluses of key-to-tape over paper tape are that it eliminates a task, it is cheap and easy to use and it is easy to restart in case of a breakdown; the minuses are that it requires twelve weeks of training, there are time delays before errors are dealt with and input verification requires a bulky printout. The job satisfaction pluses are that the device is similar to a typewriter and thus easier for individuals with typing skills to operate and that the printouts are easy to read, if well laid out; the minuses are that keypunchers cannot see what they are punching, delays for rejects are frustrating and the paper is heavy and awkward to move and store. Similar technical and satisfaction advantages and disadvantages can be listed for each type of input, processing and output equipment.

The list of alternatives was then shortened by a preliminary evaluation against the requirements, constraints, resources and objectives put forward earlier. For the mail-order firm, a short list of technical alternatives was prepared:

· Key-to-tape input, batch-processing update, paper printout;
· Paper-tape input, batch-processing update, paper printout;
· Visual display units on-line for input, batch-processing update, paper printout;
· Visual display units on-line for input, batch-processing update, microfilm output for rejects.

A somewhat different method was used to arrive at the independent list of social solutions. The solutions were variations on a few basic strategies for organizing work. The first is the most familiar: one person does one task repetitively. In the job enlargement strategy, one person does a number of tasks, increasing variety and reducing boredom. In general, however, the tasks combined in the enlargement strategy tend to be equally simple; the total job often remains low in challenge to the worker. Further, the tasks may fail to add up to a meaningful unit of work. The job enrichment strategy seeks to overcome the first of these disadvantages but not the second, by combining a variety of tasks with different skill requirements into individual jobs.

The most advanced of the job design strategies can be called self-managing or autonomous work groups. Here, the work group of six or more job holders is the focus of the design activity, not the individual job. The objective of this strategy is to combine within the scope of a single work group the set of tasks required to perform a complete activity. These may include functions normally performed by management, such as setting performance targets and monitoring progress toward them, scheduling, recruiting and training, maintenance and cleanup. (For additional descriptions of the work design philosophy of autonomous work groups, see Poza and Markus, 1980; Markus, 1983b; Cummings and Srivastra, 1977.)

Applying these strategies to the case of the mail-order firm, a number of different social alternatives were identified. First, the office could be divided into, say, five large teams, each of which dealt with input, processing and output for a subset of orders or sales agents. Team 1 might handle sales agents whose names begin with A through E, for example. Within this arrangement, two variations are possible. A person on team 1 could perform all processing steps for the A sales agents, or he or she could perform one or more processing substeps for all sales agents assigned to that team.

In another major arrangement, the office could be divided not into teams but into tasks—for example, two or three input steps, one or two processing steps and several output steps. These tasks cut across every order and sales agent processed by the office. Each person does one task, and normally everyone in the task group performs the same task. If the first task is key-to-tape data input, there may be twenty-five or thirty employees, each performing this activity.

Other alternatives can be created by using a different organizing strategy, team versus task, for input activities than for output (with the computer handling the processing steps). Thus, five input teams might be assigned by sales agents with each team member handling all input activities for a sample of agents, but the output part of the work flow might be divided into tasks, each with several identical jobs. The three basic alternatives mentioned earlier—team with individuals performing the entire process, team with individuals performing a single task and task organization—can be permuted into six alternatives, if a different strategy is used at the input and output stages.

In the case of the mail-order firms, the three basic strategies and six variations were reduced to a short list of four:

1. Office divided into teams with each employee doing all stages of work;
2. Office divided into teams that deal with all stages of the work, but with each employee doing only one task;
3. Input tasks done in teams but with each employee doing one job; output tasks done in teams with each employee doing all stages of the work;
4. Input tasks done in teams with each employee doing one job; outputs tasks done in teams with each employee doing all tasks.

At this point, the technical and social short lists were merged to generate a set of feasible sociotechnical designs for the workplace. Technical solutions with no compatible social solution and social solutions incompatible with the technical short list were rejected. Then the list of sociotechnical alternatives was ranked, and the most desirable one was selected. The preferred solution was: Visual display units for input, batch-processing update, paper printout and input jobs in teams; each employee does one task. Output tasks are done in teams; each employee does one task. The runner-up solution was: Visual display units for input batch-processing update, paper printout and input done in teams; each employee does one task. Output is done in teams; each employee does all tasks.

The chosen solution was then outlined in greater detail and compared to the original objectives, constraints and resources before a final decision was chosen. Mumford and Weir justified their recommendation (in part) as follows:

> 1. [Teams] for both input and output tasks will work very well, because we can divide all the accounts up, so that each group of girls is totally responsible for all the work on a particular set of accounts. This arrangement will enable staff to become familiar with some of the agents. . . .
>
> 3. The output section of the group will have to deal with the four main printouts from the computer:
> (a) Input and validate . . .
> (b) Credit reject . . .
> (c) Commission reject . . .
> (d) Reminder letters . . .
>
> The chosen arrangement for the output section is groups in which the staff do all stages of work. This arrangement will be well suited to the type of problem-solving work needed on the reject printouts. Each group will be responsible for the accounts of 20,000 agents, but within that, smaller subgroups could be set up. . . . The best way to organize the work might be for each subgroup to deal with the 4 printouts and the correspondence for its own agents. Each girl would deal with a printout right through. . . . Within each group, the jobs could be graded so that new entrants started on the easiest job and gradually moved on to the other jobs. . . . The first job would be input and validate, dealing with mistakes in the input data. [1979: 196–197]

UNADDRESSED ORGANIZATIONAL FEATURES

Unlike traditional system development, which guides an analyst in designing only the technical features of systems, the ETHICS method helps an analyst to specify also those aspects of social structure and culture that contribute to satisfaction on the job. Is this as far as the system design process should go? A good case can be made for answering the question with no—at least under certain circumstances.

Compensation

In the first place, the ETHICS method is based upon a model of job satisfaction that includes as one of its core concepts the balance between employees' perceived contributions to the employer and the financial rewards received and managerial controls imposed. However, the method as described stops short of guiding the analyst through a direct examination and revision of payment schemes and performance evaluation and feedback mechanisms in the organization.

The method possibly fails to address these issues because the people applying the ETHICS method will be coming from backgrounds as computer professionals with assistance from behavioral science specialists like organizational development consultants. These people may have neither the knowledge, the interest nor the legitimacy to invade the domain of personnel compensation analysts. Further, organizational policies may discourage creating a different compensation and evaluation scheme for one department, no matter how different its technology, structure, culture and systems.

Whatever the reason for this omission, the method seems the weaker for it in light of the interaction perspective on system impacts discussed in chapter 4. There it was argued that change in the direction of system design features is much more likely to the extent that other aspects of organization were similar and compatible. Returning to the example of the mail-order firm, it is unlikely that Mumford and Weir's recommended system design would work if the payment system were heavily skewed toward individual efforts rather than team performance. However, consider the authors' description of the payment scheme in force prior to their analysis and redesign:

> The jobs are graded, with ledger-well clerks, separators and file girls at the lower end, rising to accounting machinists, correspondence clerks and commission clerks at the top end of the scale. The girls are paid a flat rate according to grade, and in addition, they earn a bonus on output, that is, on the number of batches of work they complete. This increases their wages by about one-third. The supervisors are also paid a flat rate and their bonus dependent on the output of the girls under them. [1979: 134–135]

This payment scheme is definitely heavily weighted toward individual efforts. In its current form, it does not support the teamwork design proposed and may hinder its effectiveness.

The general principle of designing compatible systems and compensation schemes has been well stated by Cherns:

> The systems of social support should be designed so as to reinforce the behaviors which the organization structure [read system] is designed to elicit. If, for example, the organization is designed on the basis of group or team operation with team responsibility, a payment scheme incorporating individual members would be incongruent with these objectives. Not only payment systems, but systems of selection, training, conflict resolution, work measurement, performance assessment, time-keeping, leave allocation, promotion, and separation can all reinforce or contradict the behaviors

> which are desired. This is to say that management philosophy should be
> consistent and that management's actions should be consistent with its
> expressed philosophy. [1976: 790]

Thus, Cherns is advocating an integrated, thought-out approach rather than a
haphazard resolution of an existing payment plan.

The antithesis of this approach can be seen in an implementation tactic
described by Locke (1980). Designers from a famous school of public health
had developed a system for use in the affiliated hospital. Called HYDRA
(History Yielder for Data Research and Analysis), the system collected and
analyzed data about patients undergoing cardiac catheterization, a diagnostic
procedure for heart disease. In order for the system to work, someone had to
fill out the data entry forms, and the hospital's residents (junior cardiologists)
were designated. These residents rotated through the department so quickly
that they often did not see the results or the usefulness of HYDRA; consequently, filling out the forms had a very low priority in their already busy
days. The designers eventually decided to remedy the problem by paying the
residents a small sum of money for each data form they filled out. This tactic
may have capitalized on the relatively low pay typical for residents, but it
could hardly have been more at odds with the ethos of better patient care.
While it may have temporarily alleviated one data collection problem, it failed
to ensure the quality of the data and did nothing to enhance the use of the
system's outputs. In a word, systems and payments should be compatible, but
compatibility cannot be grafted onto an incompatible situation.

Structural and Political Features

A second concern with the ETHICS method is also rooted in the interaction
perspective. There it was argued that an implementor can reduce resistance
and increase the chances of successful system implementation by matching
the system to the existing organization. If, however, the organization does not
currently meet the managers' objectives for it, then both system and organization must be designed to be compatible with the targeted outcomes. Sometimes the goals are technical, like improving the throughput of the work flow;
sometimes they are social or cultural, like improving job satisfaction or
turnover; but sometimes the goals are political, like ensuring that the different
subunits of the organization work to achieve the same ends.

In light of these concepts, the ETHICS method can be seen to produce
a match between the work flow and the system in a department while
achieving designers' technical and social (improved job satisfaction) objectives. The method does not address interdepartmental functioning, however,

and thus cannot help to specify a match between the system and the structural and political aspects of the organization.

Mumford and Weir appear to be aware of this limitation on their otherwise excellent approach. While discussing self-managing or autonomous work teams, which they identify as the preferred design concept for the social aspects of intradepartmental work, they note that several conditions must exist before the concept would work:

> The self-managing group can be excellent in the right situation and it provides a stimulating work environment in which staff can readily develop their talents. However, for it to succeed certain things are necessary. First of all the work of the department into which the self-managing groups are introduced must provide scope for multi-skilled work that provides challenge and responsibility. In many situations the work has been so strictly allocated between departments that no rearrangement of tasks or creation of self-managing groups can make it much more interesting. In this kind of situation, the challenging, problem-solving aspect of the work has been separated off and handed over to a specialist group in a separate department. Any real improvement in work interest can now only be achieved if several departments are merged together, thus providing the required work variety. [1979: 32–33]

In short, an analysis and design method based on the interaction approach must address not only the basis on which departments are structured and interrelated but also the work flow within them.

CONCLUSION

Traditional system building is characterized by three design features:

1. System design features fully specified before execution;
2. Emphasis on technical features to the exclusion of social, structural and political features;
3. Professional control over user participation.

The first feature assumes that designers and users know beforehand everything they will want to do with the system. The second assumes that organizational changes are caused by technology rather than by the interaction of technology with its context. The third assumes that the legitimate role of computer professionals is to manage and control the contributions of users to system building.

New technologies for system building are beginning to challenge these assumptions. Evolutionary design and minimum critical specification attack the principle of full specification. Both assume that organizational change and learning about systems take time. Evolutionary design builds time for learning and change into the system-building process while maintaining the computer professional in the role of managing and controlling user participation. Minimum critical specification assigns this role to the users.

Evolutionary development and minimum critical specification are not widely used as alternatives to full specification. Evolutionary development is becoming increasingly popular, at least in academic circles, but it has been prescribed only for systems with certain technological features—for example, decision support systems employing interactive computing, data base management and user-friendly inquiry languages. Because most users of decision support systems are mid- to high-level managers, these innovative techniques are effectively restricted to the upper echelon in the organization.

For the routine jobs of low-level employees such as clerical staff, blue collar workers and line supervisors, only full specification techniques are considered. However, the jobs of these people are already overspecified in many cases. Consequently, introducing new systems often squeezes out the last vestiges of autonomy and discretion, stifling creativity and flexibility. It has long been observed that workers can paralyze an organization by observing rules too closely. The philosophy of full specification dooms implementors to developing rules that will produce desired outcomes when obeyed to the letter, not to the spirit. If effectiveness requires that employees understand and observe the spirit of the rules, evolutionary design and minimum critical specification may be suitable techniques, regardless of system type or user rank.

The ETHICS method attacks the assumptions that change is caused by technology and that the computer professional must control user participation. The method assumes that both the technical system and the social system must be designed and that they must be designed interdependently. The proper role of the computer professional is to develop and evaluate alternative computer system designs; users develop and evaluate alternative social system designs. Professionals and users together can participate in the matching of social and technical alternatives.

The interaction perspective on system impacts highlights additional features of organization not addressed by the ETHICS method. These include other managerial systems like compensation schemes and human resources management policies and organizational structures and politics. A system building process grounded in the interaction perspective must address these issues, which is the task of chapter 8.

Chapters 6 and 7 continue the exploration of the system building process. Whereas chapter 5 has examined the design features of system building, the next two chapters examine the context in which system building takes place and with which system building design features interact, resulting in the systems implemented and used.

Users and Vendors

INTRODUCTION

A major source of the design features of systems today is the computer world—the professionals and firms who provide computing equipment and services. They may provide complete, ready-made systems of any of the five types—operational, monitoring and control, planning and decision, communication and interorganizational—that are purchased by organizations and used relatively intact. Or they may offer assistance of various types to organizations who would rather build systems predominantly with internal professional and technical resources.

Each vendor offers a unique package of skills and resources, equipment, systems, methods and services for sale to the organizations that use systems. Simply because of their structural position with respect to system-using organizations—that is, located in a different firm and financially dependent upon sales—vendors have interests and objectives that vary at least slightly from those of system-using firms. For example, the vendor may wish to make a profit from a quick sale with little follow-up service, whereas the system-using organization may desire to obtain cheaply the expensive talent required to install and maintain the system and train the people who will use it.

Consequently, the process of buying or building a system with the help of a vendor is a political situation in which each party has the opportunity to achieve its goals or objectives at the expense of others. Ideally, both may come

122

away from the exchange satisfied, but the potential for winners and losers clearly exists. In many cases, the buying or building of the system is the battleground, and its specific design features will reflect the interests of the winner or the uneasy truce among the conflicting parties.

The tensions of the buyer/seller relationship alone are insufficient to explain the nature and extent of the influence that vendors exert over the design features of a system, however. An additional factor is the complexity and variety of the computing world (Kling and Gerson, 1977; 1978). The computing world is populated by a large number of vendors of different types, each of whom has interests that differ from the others. In addition, since many acquisitions of systems involve more than one type of vendor, the design features of the system eventually installed may reflect the balance of power among vendors as much as the balance of power between vendors and users.

THE VENDORS

Unfortunately, the computing world is not organized around systems, as categorized in chapter 2. The computing world is organized around the technology of computing as it has evolved from the 1940s to the present day. Consequently, an understanding of the computing world requires at least rudimentary knowledge of computing technology.

A system that entails computer or communications technology consists of two components: the computer (and/or telecommunications) component and the application component. The distinction between computer and application components can probably best be made by analogy to a player piano: The application is the music roll that plays on the computer system piano. Applications perform what a user (listener) can perceive (hear) as useful work (music). The computer component is the engine or machine that drives the application and enables it to produce the outcome desired by the user. Just as both the music rolls and the piano will affect the quality of the sound heard by a listener, characteristics of both the application and the computer components will affect what a computer user can and cannot do with a system.

It is traditional in the computing world to refer to the computer component of a system as "the computer *system*," or simply, "the system." This usage reflects the fact that the computer is only a small part of the total package of equipment and software required to make the computer perform a useful application. What the computing world calls "the application" is closer in meaning to what has been called "the system" in this book. The two are used interchangeably. The technological package of computing components is referred to as "the *computer* system."

People sometimes try to express the difference between the computer system and the application as the difference between computer hardware (e.g., central-processing units, peripheral devices, telecommunication lines) and software (programs of instructions written in a computer language for execution on the central-processing unit). While this may be an aid to understanding, it is not necessarily correct and may eventually be misleading, for several reasons.

First, a large part of the computer system consists of software. Each of the several digital devices in a computer system may be controlled by software programs that are quite distinct from application programs. In addition, special programs, called "systems programs," are frequently written to coordinate the interaction of the various hardware components in the computer system. Second, the distinction between hardware and software is rapidly eroding because of technological developments. Halfway between hardware and software lies firmware,* also called "microcode," which is being used increasingly both to control individual devices and to develop parts of the systems programs. Some experts anticipate that parts of application programs may ultimately also be implemented in firmware. Finally, certain systems programs are being replaced by special function hardware devices like front-end computers (for handling telecommunications processing) and back-end computers (for handling data base processing). Thus, the hardware/software distinction is not an entirely appropriate way to define the difference between the computer system and the application.

The major components of a computer system are the following: one or more computers (also called processors, each consisting of a central processing unit and main memory for instructions and data), secondary storage devices for programs and data, input/output devices, cables and/or telecommunications lines and networks to interconnect these devices and systems software (to coordinate the operation of the interconnected devices, to control the movement of data back and forth between secondary storage and the computer and to control the flow of data between input/output devices and the processors).

Applications, today, are almost always software programs. They have functions that a manager can readily understand—for example, budgeting, production scheduling, inventory control, text processing or voice-store-and-forward. These applications run or operate on computer and/or communication systems to produce the output reports and capabilities that support operations, control, planning and decision making and communication in the organization.

---

*Firmware performs computations more slowly than hardware but faster than software. Developing firmware is cheaper than developing hardware but more expensive than developing software. Changing firmware is easier than changing hardware but harder than changing software.

Types of Vendors

The vendors of computing systems group roughly around these two basic components, the computer system and applications, largely for historical reasons. Appendix A describes the evolution of the computer world and the different types of firms that populate it. The different vendor types are summarized in figure 6.1.

Strategies for Acquiring Systems

Clearly, not all organizations that use systems find it necessary to deal with every type of vendor listed in figure 6.1. Few would find it desirable because of the amount of coordination required to manage them. Over time, the relations between users and vendors have settled into a few basic patterns that specify how many vendors of which type the system-using organization will have to deal with. These patterns can be described in terms of strategies for acquiring systems.

There are three basic strategies for acquiring the application component of a system that involves computers. The strategies may be called "in-house development," "external purchase" or "do-it-yourself computing," although these terms oversimplify what is involved in each case. The in-house development strategy assumes that the system-using organization already has a computing infrastructure of professionals who are able to develop software. External vendors are used to develop and help maintain the computer system only. The external purchase strategy relies on vendors to develop software and possibly also to own or operate the computing system on which it runs. The do-it-yourself strategy avoids an internal computing department altogether, relying on external vendors to supply the computer system and general purpose software, which the ultimate user may tailor to personal needs by simple operations. An example is the person who uses a personal computer for text editing.

In-house Development. In the early days of computing, in-house development was the only alternative available to those organizations that wished to explore the commercial potential of computers. Because there was no software or services market to speak of and almost no trained computer professionals, the pioneering companies found themselves with a big piece of iron and they had to learn what to do with it. Gradually, they developed in-house programming expertise, aided by the growth of a labor market of mobile professionals. Until the 1970s, it was common for all but the smallest computer-using organizations (e.g., hospitals) to design and program most of their applications internally. Over time, the number of people dedicated to

FIGURE 6.1
Types of Vendors

Vendors of Computing Systems	Descriptions
Hardware Manufacturers	
Component Manufacturers	Build one or more of the component parts necessary for computing capability, central-processing units or peripherals—e.g., storage technology
Integrated system manufacturers	Build or resell all components necessary for providing computing capability
Mainframe computing system vendors	e.g., IBM
Minicomputer system vendors	e.g., Digital Equipment Corporation
Microcomputer system vendors	e.g., Apple
Plug-compatible manufacturers	Build one or more components that substitute for the products of a dominant vendor—e.g., Amdahl
Leasing Companies	Provide financing, insurance and other services for purchasers of expensive computing equipment
The Telecommunications Sector	Provide the equipment and services that allow people and computing equipment to communicate over long distances
Publicly owned network operators	e.g., AT&T
Specialized common carriers	e.g., MCI
Private networks	e.g., ARINC in the airline industry
Computer System Marketers	Sell equipment, service and possibly software for mini- and microcomputers
Licensed dealers	
Retail stores	
Mail-order firms	

FIGURE 6.1
(Cont.)

Vendors of Applications	Descriptions
Processing Services Firms Batch-computing services On-line computing services	Supply raw computer power, access to specialized data banks or data-processing services via proprietary applications programs.
Software Houses	Build and/or sell software packages
System Houses and Original Equipment Manufacturers (OEMs)	Produce integrated packages of computing equipment and applications software
The Professional Services Sector	Provide custom programming services, facilities management and consulting services

developing and maintaining computerized applications grew to the point where the organizational budgets for information services could be measured as several percent of sales.

Developing applications in-house certainly does not preclude the use of outside processing services. However, most companies with sizeable internal development efforts usually run most of their applications on internally owned and operated computers. This is partly due to historical reasons since the computer usually preceded software into organizations and partly for reasons of political clout since a large and gleaming glass-walled computing center is tangible justification for a large budget and an army of people to program and tend the machines.

There are two drawbacks to in-house development. First, it does not allow an organization to avoid all contact with external vendors. Organizations must still acquire computing systems or processing power (as well as computer professionals) from one or more outside sources. Second, this way of acquiring a system can carry a high price tag. It requires the development and ongoing maintenance of an organizational infrastructure (a structure within the larger structure) that usually has much greater capacity than that needed to solve a single system need. Therefore, once created, the infrastructure must ensure its survival, the implications of which are discussed in chapter 7.

External Purchase. The external purchase alternative for acquiring a system has been steadily growing in importance since the mid-1960s. As the cost of hardware decreased dramatically, it became clear that the major computing expense is the labor of professional programmers. Purchasing externally developed solutions cuts this expense. External purchase comes in four varieties. Two of them, processing services and turnkey applications, are intended for those who wish to avoid the hassles, expense and infrastructure of large-scale general purpose computation. Two others, packaged software and custom contract programming, are intended for those who have already made the investment or the decision to invest in an infrastructure. The first two varieties reduce the number of vendors with which an organization has to deal (around a given application) to one; the latter two increase the number of vendors by one. Whether this is a bug or a feature is a question that is addressed shortly.

Do-It-Yourself Computing. Sometimes called "personal computing" or "end-user programming," this is the most recent of the alternatives for acquiring a systems solution. Currently, it is the least used of the four. However, if the rapid growth in the sales of personal computers to businesses is any indication, this may become a preferred way to solve new systems needs. Successful do-it-yourself computing requires three things: interactive computing, data and easy-to-use, general purpose applications software. Interactive computing may be provided by a personal computer or by terminal access to interactive processing on a system maintained either in-house or by an outside vendor. Data may be abstracted from a corporate data base, supplied by an external data base vendor or entered manually after having been collected from any sort of source (including one's imagination). Easy-to-use software may include very high-level programming languages that do not intimidate managers, general purpose packages like Visicalc which can be used after brief self-paced instruction or applications generators that are skeleton application packages (e.g., accounts payable programs) that can be tailored to particular needs by specifying certain parameters (i.e., filling in the blanks).

To the surprise of many people, do-it-yourself computing has received a very enthusiastic response from many managers who formerly had little use or regard for the centralized information services in their own organizations. It appears that a major reason for the popularity of this alternative among users is that it reduces their dependence on system professionals. At the same time, it increases users' control over the timing of development, operation and changes in systems (Rockart and Flannery, 1981). However, while the users may value these features of end-user computing, many system professionals see them as bugs that threaten their importance and role in system-using organizations. Figure 6.2 summarizes the number and types of vendors involved in each strategy for acquiring a system.

USER-VENDOR DYNAMICS

The interactions of vendors and users take the shape they do because users' interests differ from those of vendors and because the many vendors that may be involved in a single system may also have conflicting interests.

Users' Interests

The interests of users in their dealings with vendors can be described as follows: Users desire the ability to take advantage of improvements in the cost

FIGURE 6.2
System Acquisition Strategies and Vendors

Strategies for Acquiring a System	Vendors Involved
In-house Development	
Single-vendor policy	One vendor—the integrated systems manufacturer and internal department for software development.
No single-vendor policy	Multiple vendors for equipment, services, software; internal department for software development.
External Purchase	
Processing services	One vendor, a processing services firm.
Turnkey applications	One vendor, a systems house.
Packaged software	At least two vendors, one for equipment, one for software.
Custom contract programming	At least two vendors, one for equipment, one for software.
Do-It-Yourself Computing	One vendor for equipment or for processing services.

or capabilities of computing without having to make expensive adjustments or conversions. The biggest conversions occur when a user of general computing with a large portfolio of applications switches from one vendor of computer systems to another. Because there is so little standardization among vendors in most types of computing equipment and software, and because applications are often sensitive to the computing systems on which they run, switching vendors may entail expensive modifications of application programs. Even changes that seem relatively minor can be very expensive in time and money. For example, a major insurance company recently announced a conversion from one dialect of COBOL, a programming language, to another version of the same language (with no change in the computing system) and estimated the cost at $20 million (Shoor, 1981).

That these costs may deter companies from changing vendors can easily be imagined. Imagine someone who bought a Beta video recorder for home use three years ago and who since then has accumulated a library of cassettes worth $7,000. In the process of replacing the machine, this person may discover that VHS recorders are more popular in that geographic region, offering a wider selection of models, service centers and recorded programs than Beta. A new VHS recorder is only $1,000, but it won't play the $7,000 worth of Beta cassettes. What to do? Most people in this situation would buy another Beta.

Computer users behave the same way. It has been estimated that IBM gets only 10 percent, and Honeywell only 25 percent, of their mainframe computer business from new customers. The majority of their revenues comes from repeat business. In general, when computer systems are operating satisfactorily and appear likely to continue to do so, organizations will avoid conversions and use their market power to pressure vendors into providing smooth upgrades. [See Dolotta et al. (1976) for an impassioned plea from users to vendors for stable operating systems.]

This implies that organizations value stability in the vendors that supply them because, if a vendor goes out of business, the customer either takes over all further maintenance and foregoes enhancements and upgrades or converts to another vendor's product. Unfortunately, vendor default is, if not common, at least familiar in the computing world. For example, General Electric, RCA, and ITEL among others have pulled out of the computer-manufacturing business and have had to make arrangements with other computer companies for the continued maintenance and support of their installed base of computers. In addition, while Honeywell, Univac, and National Advanced Systems, respectively, took over these functions, users rightly had considerable concern about the depth of knowledge of the new field service personnel, how long the new support agreements would last and the now-inevitable conversion to a new product line when existing capacity was exceeded or support discontinued.

Vendor stability is more than the ability to remain in business and to provide a line of compatible products that allows customers to upgrade gracefully as their needs change. It is also the ability to deliver on their promises to customers. Not all of the vendors who fail to deliver are new or fly-by-night; even the best respected names will be associated with a disaster at one time or another.

Computing disasters may spell bankruptcy for the small firm. *The New York Times* reported that Triangle Underwriters, a forty-four-year-old firm, went out of business as the result of "a multiplying series of computer errors" involving Honeywell equipment. The company has sued that vendor and was awarded "$2 million in damages and interest on the ground that Honeywell fraudulently misrepresented the [H-100's] capability and, in effect, destroyed Triangle's $5 million business" (Schuyter, 1980).

Another small business has made even more extensive waves as a result of an unhappy experience with Burroughs. Quality Books is still in business, but as a result of problems with the B-800, the firm "lost business that will take years to recoup" (Apcar, 1980a). The company filed suit against Burroughs for lost business damages. In the process of collecting data, lawyers talked with other companies that had bought the small business computer and discovered that Quality Books was not alone in their problems. Quality Books placed an ad in the *Wall Street Journal* to solicit complaints from other customers. As a result, at one time, the attorney for Quality Books was representing thirty court actions against the vendor ("Burroughs Found Guilty of Fraud," July 7, 1981). Burroughs struck back with a countersuit against Quality Books for the loss of its business caused by the bad publicity; it also launched a massive information campaign that used videotaped testimonial from satisfied customers (Apcar, 1980b). However, Burroughs lost a round when several users of its small business systems, the B-80, B-700 and B-800, were awarded a total of $750,000 in punitive and compensatory damages.

> The suits have created a stir in the computer industry because they attack a bread-and-butter issue: to what extent, if any, are computer makers financially liable for the business costs and damages caused by faulty machines and systems? ["Burroughs Found Guilty of Fraud," July 7, 1981]

Whatever the answer to this question, damages are small recompense to businesses that suffer or fail because of misplaced confidence in a vendor.

In short, the interests of users with respect to vendors stem from the dependence that users develop on them. When an organization acquires a computer system, it enters an ongoing and expensive relationship with a vendor. The consequences of a bad choice—a vendor that goes out of business, one that provides poor customer support or one whose product line lacks an ability for smooth upgrading—are harsh. The consequences of a good choice—for example, a vendor who will tailor a general system to customer needs—may be more pleasing but may still lock the customer into a relationship of dependence.

Vendors' Interests

For their part, vendors need to ensure continuing revenues. When they can't get these from new customers (i.e., when the industry is mature and the market saturated), they must generate an ongoing stream of improvements in price performance and drop from their product lines items that no longer produce sufficient revenue. Every few years, vendors of hardware will bring out new models with larger memories, faster access times, new features or better price performance (read reduced price, bigger and better or both). In the software game, this strategy takes the twist of staged releases, with periodic reissues that fix bugs found in earlier versions, add new features or improve operating efficiency. As equipment and software releases age beyond their useful economic life (to the vendor), the vendors will refuse to maintain and enhance them.

The prospect of having to undertake maintenance in-house is usually sufficient to goad most system-using organizations into keeping relatively up-to-date, even if doing so entails a fair number of hassles. Part of the reason for this is that users often find small, incremental changes easier to take than one big conversion, and vendors know this. Another reason is that computing organizations (vendors and users alike) fear the reputation of being obsolete. Being obsolete makes it harder for vendors and customers to attract qualified computer professionals, who measure their career mobility in terms of their experience with the latest technology.

These factors conspire to produce almost continual change in the computing world. However, not all of the dynamics of the computer world lead to change unwanted by the user; several structural features of the computer world serve to inhibit changes that users would like to see take place. Individual users often identify needs they believe could be solved through the technological capabilities of various vendors yet are unable to interest vendors in building something to meet these needs.

Many factors contribute to this state of affairs. Not least among these are vendors' perceptions of the size of the market for a particular computing innovation. Unless the vendor believes that a large enough market exists so that development costs can be recouped and a profit made, that vendor is unlikely to invest the resources necessary to build the product.* Second, a vendor may already have developed an innovation but choose not to market it (immediately or ever) for a variety of reasons (Gilchrest, 1980). For example, vendors must consider the economics of existing equipment throughout the production life cycle. If marketing a technical innovation would cause the premature obsolescence of an existing product, the vendor is unlikely to market it.

Other considerations include the behavior of dominant vendors, the attitudes of bellwether customers and fears about the attitudes of rank-and-file customers. Certain innovations in computer systems have been developed many years before their widespread commercial acceptance. Their developers were unable to market them successfully until they were adopted and marketed by a dominant vendor. An example is virtual memory, developed in the 1950s, but not a common feature of computer design until the 1970s when its discovery was announced by IBM (Gilchrest, 1980).

Even dominant vendors will sometimes test market innovations with favored customers, whose response can either be expected to reflect that of others or whose adoption (or failure to adopt) could influence other customers. If these bellwether customers react coolly, a vendor may drop an innovation from further consideration. Finally, vendors may simply choose not to promote innovations that entail conversions for fear that customers would react negatively. This is probably what happened in the case of IBM's never-announced Future System computer. For several years, IBM worked to develop a computer that was architecturally different than the S360-370 series. The new architecture promised many benefits but might have required conversion of application programs written for the other system. Although the Future System was never formally announced, news of its development leaked out and was greeted with apprehension by customers. IBM scrapped the project in 1975, and its large computer systems announced through 1980 represented a relatively smooth extension of the S360-370 architecture.

In general, then, there are almost as many forces for stability in the computer world as there are forces for change. Unfortunately, however, from the users' point of view, stability often prevents desired change, while innovation often promotes undesired change. While users do have some options for successful negotiation with vendors, some observers believe the contest is quite uneven, with the edge to the vendors (Kling and Gerson, 1977; 1978).

---

*IBM was rather late entering the computer manufacturing business because it did not perceive adequate commercial potential for the device.

Differences among Vendors

The preceding discussion may have implied that the vendors of computing technology and services share equally in those interests that differentiate them from users. In fact, however, the interests of vendors can vary considerably, in part due to the sheer numbers and diversity of participants in the computer world. The variance in vendors' interests can be illustrated with a few examples.

In the software packages marketplace, a vendor's interest in a customer may be short term or long term, and it may or may not extend beyond the apparent content of the immediate transaction. Some software houses choose to sell the rights to use their packages outright. They realize revenues on a one-time basis (except for what they may charge for yearly maintenance contracts) and, consequently, may view their relationship with customers as a one-shot deal. Others lease their software on a monthly basis. These firms may have the security of knowing that it would be difficult and expensive for their customers to convert, yet the continuing nature of payments and the lengthy time period over which revenues are extracted from a customer may induce this type of vendor to pay continuing attention to user needs. Still other vendors view the sale of software primarily as a sideline or as an inducement for the customer to acquire other services. Thus, time-sharing firms will attempt to increase sales of computer time by renting software packages that run on their equipment. Also, consulting and certified public accounting firms sometimes advertise specialized application packages that require their expertise to adapt them to clients' circumstances. In this case, what an organization buys is not so much software as a (possibly long-term) consulting arrangement.

The vendors of mainframes and integrated systems also sell packaged software. The way their interests differ from those of the so-called independent vendors (nonmanufacturers of computers) can be seen in the public debate waged in the pages of computing trade journals. The independents charge the mainframers with a variety of unfair tactics. They argue, for instance, that the software-producing units inside hardware firms have advance knowledge of technical changes that can affect the product offerings of the independents. Further, they argue that manufacturers can exploit their advantage by deliberately modifying computer system technology in ways that make competitive products obsolete or that force the independents into extensive and expensive modifications. A counterargument points out that software products in the hardware vendors' portfolios would also be adversely affected by such moves, but this fails to allay the fears of the independents.

Other so-called unfair tactics include selling bundled solutions, preannouncements, the ease-of-installation software tie-in and the software support tie-in (Goetz, 1980). Charges of the bundled solution tactic have been leveled at IBM's 4300 computer series, which uses microcode ostensibly to improve the performance of its systems software. However, microcode may also effectively prevent independents from supplying systems software because they have to figure out what functions the microcode performs and tailor their programs to work on the microcoded computers. By treating as confidential the microcode and specifications about how different pieces of software interact with each other, vendors can give software houses an extremely difficult business environment (Goetz, 1980). In this way, vendors can force would-be competitors to wait until the product is delivered before they can begin developing an alternative. This results in the tremendous power of industry leaders like IBM to set standards that other vendors ignore at their peril. Products incompatible with the standard (even if it has not arrived on the market) face small markets and short lives.

Independent software vendors also claim that giants like IBM can ruin their business by preannouncing software products well before they are ready to deliver these to customers. Many industry watchers have observed that early announcements of new hardware by IBM may cause potential buyers to defer purchasing competing products for several years until IBM's products are available. This technique probably works for software as well. The farther in advance of delivery such preannouncements are made, the more damaging their potential effect on the competing products of independent firms.

Independents sometimes see potentially illegal tie-in tactics in efforts by hardware vendors to improve the ease of installing their software products. The widely acknowledged installation problems experienced by users of IBM's operating systems in the 1970s stimulated sales of installation aids and operating system enhancements by independent vendors. This business was threatened by IBM-developed improvements in which the various operating system components are prelinked, pretested and delivered on one magnetic tape. "But, to receive this tape, the user must take IBM's operating system and five or more of its program products. . . . [This] appears to be a classical tie-in" (Goetz, 1980).*

Another way in which hardware vendors like IBM can make life difficult for independent software vendors is the software support tie-in.

> Another new IBM offering is in the area of "operating system support". IBM now charges for such support but requires that the user have an IBM

---

*The *Computerworld* article goes on to point out that these changes, together with the use of microcode, create problems for plug-compatible computer companies as well as software independents. All sectors in the shadow of the mainstream vendors fear the consequences of their actions.

> operating system for such . . . support. Obviously, software product com-
> panies and compatible CPU companies are at a distinct disadvantage.
> Software companies producing compatible operating system extensions will
> find users reluctant to buy their products because of IBM's support tie-in.
> Similarly, CPU-compatible companies will find users reluctant to buy their
> CPU's. (Goetz, 1980)

The essence of the support tie-in is the ability of a hardware vendor to deny maintenance to users who choose operating software from other sources or to refuse operating system maintenance and training to users who choose hardware from another source. The legality of these moves has always been questionable: Data General, for example, lost an antitrust suit to Fairchild Semiconductor for requiring customers who wished to acquire its systems software to purchase its microcomputers (Bulkeley, 1981). In addition, in mid-1981, IBM agreed under legal pressure to provide support services for its software to customers who purchased plug-compatible mainframes (Chase, 1981). The interests of the hardware vendors are such that they are motivated to try tactics like these.

Similar public debates, revealing underlying differences in vendor interests, have also occurred in the processing services segment of the computer world. The Association of Data Processing Service Organizations (ADAPSO) has spearheaded efforts to prevent communications common carriers like AT&T and large banks like Citibank from participating in the processing services industry on the basis of alleged unfair competition. ADAPSO claims that the carriers and banks can subsidize their data-processing services with revenues from other lines of business, thereby unfairly setting prices below costs. At the same time, other firms in the services sector must purchase communications services from the common carriers, who may charge too much, cause delays or provide inferior service to dependent competitors. Large banks, ADAPSO fears, may give away data-processing services as an inducement to their banking customers and thereby threaten the revenues of firms whose sole business is computer-processing services.

Implications of Vendor Differences

There are several consequences of this diversity of vendor interests for system-using organizations. In the first place, any computing solution involving more than one external vendor is certain to entail significant problems. This is due at least partly to the general lack of compatibility among the equipment and services of different vendors. But much of the reason is due to the lack of common objectives, even among vendors whose products may work together rather well. This lack of common objectives can lead to conflict rather than cooperation on joint projects. Many computer center managers have witnessed, to their chagrin, the finger pointing and name calling that can occur when trouble arises in a multivendor project.

After a few experiences of this sort, many organizations adopt a single-vendor policy. For those that do application development themselves, the single-vendor policy means restricting the purchase of computing equipment and systems software to a single integrated systems manufacturer. When a data center manager says that his or hers is an all-IBM shop, he or she is referring to the results of a single-vendor policy. For those organizations that prefer to purchase total computing solutions directly from an external vendor, both processing services and turnkey applications offer the benefits of one-stop shopping with a single vendor.

The downside of the single-vendor policy is that, in making the decision to go with one vendor, an organization gives away a great deal of power to, and in turn becomes quite dependent upon, that sole source of products or services. This is a position that few organizations like and that few large organizations tolerate. It is common practice for assemblers of electronic products (like computers) to refuse to design a specialized component or chip into their products until they are assured that a second source for that part exists. They do not want to find themselves unable to deliver a product because their sole supplier has gone out of business or failed to provide the volume of parts promised. Similarly, a user of computing goods and services could wind up vulnerable or helpless if its only vendor were to go out of business, fail to deliver promised software or to discontinue maintenance on a product in which the user has heavily invested.

A second consequence of the diversity of vendors is that some vendors will be able to supply better solutions than others to organizational needs. Ironically, the better a vendor meets a firm's needs, the more likely the firm is to become dependent on the vendor's package of goods and services and the more difficulty it will have switching to another vendor. Obviously, users may reduce dependence on vendors in many ways. For one thing, they may join consortia of other similar users and contract with vendors to produce a system for all. This arrangement gives the users more market power in dealing with the vendors because it increases the size of the contract and, hence, the profit potential to the vendor. Under these conditions, vendors are likely to be more responsive to user requests for fear of losing the job. Legal contracts and lawsuits can also help to ensure that vendors deliver on their promises. Users groups are still another way that users can exert pressure on vendors. User groups are sometimes established by vendors to obtain market information or to decrease customer dissatisfaction, and they are sometimes established by users to communicate needs, desires and problems.

Of all the tactics for dealing with external vendors, establishing the internal capability to evaluate and monitor vendor performance is the most common. However, building the system infrastructure is expensive and introduces its own complexities and constraints, as chapter 7 shows.

A third consequence of the diversity of vendor interests is perhaps the most significant for system-using organizations. When multiple vendors are involved in the acquisition of a system, their different interests create a political situation. The result of their negotiations is a system with certain design features that may achieve neither the intentions of those who commissioned the system nor the best technical skills of the vendors and computer professionals.

These implications of vendor diversity and the dynamics of user-vendor relationships are neatly illustrated in the following true story of Masada Hospital (not its real name).

THE CASE OF MASADA HOSPITAL

In 1976, Masada Hospital began to re-evaluate its computer capabilities.* The hospital's executive director, Ben Carling, desired financial planning information not provided by existing information systems; he asked his assistant executive director of fiscal services, Don Checking, to whom the director of data processing reported, to look into the matter. Checking talked with other department heads and discovered that several of them felt the need for additional data-processing support: Medical records wished to expand its processing; admitting wanted an on-line system; the director of pathology wanted to install a standalone computer system costing almost $300,000. Bob Rawls, director of data processing, reported to Checking that, by the middle of the next year, full capacity of Masada's in-house UNIVAC computer would be reached, even without the development of additional systems. Further, the current machine would not support on-line applications. Masada began to consider alternatives.

The consulting arm of one of the Big Eight certified public accounting firms was engaged; together with a task force of people from Masada's fiscal services and data-processing departments, the consultants compiled a set of specifications describing the hospital's need for data-processing services and listed seven data-processing equipment, software and services vendors who were asked to bid on the proposal. The bids went out to three hardware vendors (IBM, Burroughs and UNIVAC), two national services bureaus, and two facilities management firms. With one week left to go before the close of

---

*Don Billings, Dick D'Enbeau, Josie Lindsay, Liz Norris and Terry Pim collected the data for this case in a course in organizational behavior at Case Western Reserve University in 1978.

bidding, a local service bureau that was not on the original list asked to be sent a request for bid. This firm managed to return a proposal before the deadline.

Given the diversity of the firms to which requests for bids were sent, the proposals submitted varied significantly in scope and detail. The hardware vendors proposed to supply equipment and systems software but little specialized hospital applications software. In two out of three cases, this approach would have required the hospital to convert or reprogram its applications. The national service bureaus had developed hospital applications software that they ran on their own computer hardware. In this approach, the hospital would have provided its data to the service bureau, either in hard-copy form via courier to be keypunched by the bureau or in electronic form via terminals on hospital premises. The facilities management firms would have hired a data-processing manager and employees, placed these in the hospital's computer department and taken over from the hospital all new system development work.

The task of evaluating such diverse proposals was not easy. An elaborate set of criteria was generated by the consulting firm to be used for scoring the vendors' proposals, and one, two or three points were assigned to the criteria according to their perceived importance. Some of the criteria used were the size, complexity, on-line capability and features of the proposed computer's operating system; the language and documentation of the applications packages provided and whether they required modifications that could or could not be performed in-house; the extensiveness of support, maintenance and training on hardware and software; cost, delivery time and vendor reliability; whether the system had the memory capacity to handle both existing and planned applications and whether additional add-on memory and peripherals were obtainable from the original manufacturer or from plug-compatible vendors.

The Big Eight consulting team and Masada's in-house task force independently evaluated the proposals using the same point-scoring method. Although their rankings were close, Burroughs topped the consultants' list while the in-house team favored UNIVAC. Nevertheless, in early 1978, Masada signed a contract with Hospital Shared Systems, Inc. (HSS), the local service bureau that had been passed over in the original request for bid.

Here is the chain of events leading up to this unexpected outcome.

Hiring the Consultant

After Carling, the executive director, had asked Checking, the assistant director, about the capability of Masada's computer system to meet future needs, Checking had given Rawls, the director of data processing, the task of writing up an initial proposal making recommendations about computer expansion. Not long thereafter, Rawls presented his boss with a ready-to-sign contract for a larger UNIVAC machine. Dismayed, Checking had asked for a report explaining how this new machine would fill Masada's undocumented needs for expanded data processing. Rawls pleaded the inability to write up such a document, leaving the task in the assistant director's hands.

Up to this point, Checking had never had to question the capabilities of his subordinates in data processing. Rawls, sixty years old, had started the data-processing department at Masada by himself in 1968, when a UNIVAC 9200 was installed. Although he had never received a college degree, Rawls had been able to take this computer system through a memory size upgrade and a major operating system conversion to handle secondary storage on disks (from tape). He had hired two additional programmers, and over the years, these three people had developed a library of over 300 programs to meet Masada's needs for software. Compared to other hospitals of comparable size (Masada had 440 beds), the cost per patient day of data-processing services at Masada was substantially lower, in spite of its extensive program library.

Checking reluctantly began to prepare a report for the hospital's executive director, describing the need for the computer upgrade based on the information Rawls had given him. Troubled by his lack of expertise in data processing, Checking had tried to solicit advice from colleagues at other hospitals. They had advised him that data base management systems were considered essential to providing the kind of integrated information desired by Carling; a talk with Rawls revealed that this software was not available for the proposed UNIVAC machine. Rawls, however, believed very strongly that the hospital's programs could be tied together without the complex data base software. When Checking expressed his growing reservations to Carling, the executive director suggested seeking a second opinion.

Checking was naturally hesitant to do so. His three subordinates in data processing believed without question that the upgrade was the right way to go: his insistence on consulting outsiders would certainly be viewed as a lack of faith in their judgment and ability. As anticipated, Rawls was visibly upset when told that the proposal had not been automatically approved. Sincerely hoping that consultants would merely rubber stamp the upgrade Rawls had proposed, Checking hired Big Eight consultants to review the proposal.

Big Eight was the only firm considered for this job. The hospital's books were audited by Big Eight; the firm had helped with the initial acquisition of the UNIVAC machine in 1968. Several members of Big Eight's consulting staff had long-term relationships with Masada, although in an audit rather than a data-processing capacity. It took Carling and Checking five minutes to approve Big Eight's proposal to do a complete study of the data-processing system at Masada.

Request for Quotation

The consultants began by questioning Rawls and his staff. It did not take them very long to find what they believed was a significant flaw in the internally generated proposal for an upgraded UNIVAC computer. UNIVAC did not offer specialized applications programs tailored for the hospital industry; in contrast, Big Eight was aware that another hardware vendor, Burroughs, did have software for several of the applications the hospital planned to develop in the future. Big Eight argued that the cost of building the software would be much greater than buying it. They presented this evidence to Checking and Carling, discrediting the internal proposal. Checking and Carling asked Big Eight to develop a set of specifications and to issue to appropriate vendors a request for quotations on the specifications. The consultants agreed, indicating that the project would take from five and one-half to nine months. Rawls expressed great concern that this large study, plus the time required to change computers, would considerably set back computer operations at Masada.

The consultants held a meeting with all administrative people in the hospital, explaining the purpose and scope of their study. They interviewed each department and division head to gather data, redoing, Rawls felt, the work that the internal team had done. Big Eight developed a long-range plan for data processing that Checking then incorporated into a budget proposal for the hospital's finance committee. After this, the plan was sent to the board of trustees for final approval, which was granted. This was the first time in the ten years since the introduction of computing at Masada that explicit long-range planning for computer processing was performed. Previously, data processing had essentially been left to run on its own.

Following board approval to proceed, Big Eight sent requests for quotation to seven vendors and met with them to discuss the specifications. Rawls continued to favor UNIVAC, and the consultants felt confident that Burroughs would make the best proposal; Checking expected IBM to outbid the others. No one really thought that the facilities managers or service bureaus would make a good showing in the bids.

Hospital Shared Systems, Inc.

While the responses were coming in from the vendors, Hospital Shared Systems, Inc. (HSS), a service bureau for several local hospitals, asked

Masada to send them a request for bid. Checking was indifferent to their request until he and several board members started receiving phone calls from administrators at the hospitals now served by HSS, asking Masada to allow HSS to bid. These political tactics infuriated Checking, who half resolved on principle to ignore the pressure. However, he compromised by phoning some of his friends at the local chapter of the hospital association, asking for the story on HSS.

He discovered that HSS had had an unusual history. Years before, a third-party insurance firm had developed a set of systems for hospital data processing that it provided in a service bureau arrangement to several hospitals in the locality. The service was widely considered a disaster, and many of the original participants abandoned it. The insurance firm found the venture an embarrassment and was happy to divest it in 1977 to a local entrepreneur named Eastwood who had made a small fortune in his own consulting business. Nationally known for his expertise in labor relations and data processing, Eastwood had the energy, visibility and integrity to turn the failing service around. He fixed and expanded the software applications and signed on nine local hospitals, which were very satisfied with the performance of HSS.

HSS was, however, at something of a turning point in its existence. Nine hospitals were too few to ensure the continued viability of the company. Further, the clients of HSS were mainly small community hospitals and did not form an impressive market base from which to attract other customers. The larger Masada Hospital would provide HSS with respectability and visibility in the marketplace, and its extensive library of internally developed applications might form the core of additional services. HSS believed it needed Masada's participation and was probably prepared to provide substantial price inducements to the hospital. The existing customers likewise recognized the advantage of joint membership with Masada and did what they could to pressure the hospital's administrators.

Checking found this intelligence interesting if not persuasive. What really caught his attention, however, was the information that his friends at the hospital association had been very dissatisfied with a data-processing consultation done for them in the recent past by Masada's chosen consultant, Big Eight. Further, it appeared that Big Eight had deliberately neglected HSS in the bidding procedure for the job at the hospital association, a fact which Checking's friend attributed to a distinct anti-HSS bias on Big Eight's part.

Checking and Carling finally decided to allow HSS to bid. In the meeting to discuss the request with HSS, the vendor raised questions that implied that the consultants had not considered several key issues. Checking's impression was that the Big Eight consultants hedged their answers. This meeting served to weaken the credibility of the consultants in the eyes of Masada's administrators. Checking said afterward that he would have hired a consultant from another city had he known anything about the situation beforehand. He considered firing the firm but balked at the thought of more time lost. It seemed unlikely that the vendor would be able to respond within the week remaining before the close of bidding, and Checking hoped they would not. He later said he had resolved not to accept a late bid from HSS under any circumstances, but his resolution proved unnecessary. HSS submitted its proposal on time.

Evaluating the Proposals

Once the bids were in, the consultants and the in-house team set out to evaluate the proposals according to the complicated point-scoring system Big Eight had developed. The scoring system was to be used to narrow the field of vendors rather than to choose the awardee. The consultants and data-processing staff independently scored the proposals and compared their results. Four vendors, including both facilities management firms, were eliminated immediately. The four remaining firms were UNIVAC, Burroughs, HSS and a national service bureau. These vendors were quite close in the point ratings, but UNIVAC topped the in-house team's list, while Burroughs was Big Eight's first choice.

Many discussions on the merits of the vendors, primarily the two hardware vendors, followed. The consultants argued that Burroughs had already developed much of the needed software; the in-house group argued against the loss of their 300 program investment in the RPG language (programs for the Burroughs machine were coded in COBOL). Some of the discussions grew quite heated, but Checking did not doubt the sincerity of either side. It seemed to him that the programmers reporting to Rawls were wavering in their support for the UNIVAC alternative but were too intimidated by their boss to speak up. The other two vendors were neglected in these meetings, despite the impressively low cost of the HSS alternative.

Realizing that the task force was deadlocked, Masada's administrators took matters into their own hands, putting aside for the time the advice of their consultants. It was clear to them that all four vendors would meet the hospital's requirements, but it was difficult to decide which would do so best. Again, they consulted colleagues from other hospitals with experience in data processing to resolve the programming language question, to determine the value of programs already developed and to learn the merits of in-house computing versus a shared service arrangement. Again, they began receiving pressure from local colleagues to try the HSS alternative.

One dilemma facing the administrators was whether and how to continue working with the employees who had participated so emotionally in the task force this far. Carling proposed a face-saving out: Since it was clear that all proposals met the hospital's technical requirements, maybe now the administrators could make the decision on the basis of overall hospital welfare. This was the strategy they finally adopted, taking care to inform Rawls and his people at each remaining step of the decision process. No one was surprised that Rawls was quite dissatisfied with this turn of events.

Carling and Checking called the administrators at the HSS hospitals and at the hospital association and told them to leave them alone; they were going to make their own decisions without interference. Then they sat down for extended discussions with HSS. At this point, HSS, which had already made a very attractive financial offer, indicated its willingness to make even more concessions. Masada's top administrators began to consider the vendor more seriously.

Negotiating a Contract

A key issue in the bargaining was the anticipated difficulty of converting to the new system, since Masada wished to continue running several of its own custom-tailored applications in the new hardware environment. Before HSS would make any commitments on this point, they wanted to examine Masada's existing software. When Checking asked Rawls to give HSS the source code (the programs in human readable form) for Masada's programs, he objected strenuously. Refusing a direct order to hand over the code, Rawls bypassed Checking and the hospital's executive director and took his case to the corporation, the legal arm of the religious order that owned the hospital.

A policy of the corporation was that real assets could not be sold without board approval. Rawls argued that the software was a real asset because it could have been marketed and that, therefore, turning over the software constituted improper behavior. Resolving this issue involved contacting the mother superior of the order, who decided that the software was a special case not covered under the title of real assets. When Rawls read the letter from the mother superior indicating her belief that software was not a real asset, he gave in and released the source code to HSS.

Carling recommended firing the data-processing manager for his insubordination, but Checking believed that this would make implementation of a new system more difficult. This was not, however, the end of their problems with Rawls.

With source code in hand, HSS made firm commitments to Masada. After much discussion, Carling decided to accept HSS as the hospital's data-processing vendor and to begin the process of negotiating contracts. Their decision criteria included these points: the lower cost of the HSS proposal, cash guarantees of performance, substantial savings of costs that would have been required to remodel the computer room for an upgrade, the security of joint membership with a group of hospitals (particularly since this might enhance chances of approval by the local health systems agency) and good public relations from the lower costs of using a shared system. They informed Rawls and his people of their decision but received no response from them.

Checking and Carling also sat down with the consultants from Big Eight and explained that they believed the hospital should go ahead with HSS. The Big Eight project leader responded that he too felt he could recommend HSS, provided HSS met certain conditions. To this, Checking replied that the consultants had little choice but to accept the hospital's decision, regardless of conditions. Big Eight would be retained to see Masada through the negotiation process, but the implication was clear: Their role was to support Checking and Carling in their presentation to the hospital's board.

By feeling out members of the board of trustees' long-range planning committee, Checking learned that a contract with HSS would be approved if the remaining issues could be resolved. Thus, meetings were held between HSS and the internal project team to work out the details of implementation. Here it was made clear that the decision did not jeopardize the jobs of Rawls and his programmers. A two-year implementation period was anticipated, at the end of which Rawls would be at retirement age. Further, the hospital needed its internal people to work with HSS to tailor applications to its individual needs. These assurances made little difference to Rawls, who staunchly maintained that the decision was a bad one. In one meeting at which feelings ran particularly high, the president of HSS invited the data-processing staff of Masada to tour his facility. The two young programmers went and were duly impressed by what was clearly more advanced technology than the hospital's. They returned quite sold on HSS. Rawls, however, found excuses not to go. Even later, after his personal tour, his acceptance was grudging at best.

The final round of negotiations with HSS involved Checking, the president of HSS and the two Big Eight consultants. The clashes between the Big Eight project leader and HSS's president were so fierce that Checking finally excluded the consultant from all further meetings. Clearly, the bad blood generated by their prior professional associations was interfering with their ability to work toward Masada's welfare. After this, Checking discussed each meeting privately with the consultant and asked his opinion. By the close of the negotiations, Masada had received 80 percent of its demands from HSS.

At the meeting with the long-range planning committee of the hospital's board of trustees, Carling and Checking made the bulk of the presentation. Big Eight gave a brief talk that fell short of actively endorsing HSS but that gave the impression of support for the decision of the hospital's administrative officers.

CONCLUSION

While the case of Masada may be an extreme example, it illustrates many of the dynamics that occur at lower levels of intensity in almost every user/vendor interaction. In the first place, there were differences of interests among the key parties to the relationship. On the user side were the hospital administrators, Carling and Checking, and the data-processing group headed by Rawls. The former desired low cost and high quality of computing services as well as a solution to computing needs that would be good over the long term. The data-processing group was divided among themselves. Rawls wanted continuing validation of his efforts for the hospital and probably feared an inability to retrain himself technically so soon before his retirement. The younger programmers may have shared his sentiments to some degree but also wanted a state-of-the-art technical environment in which to work because this would enhance their career options.

On the vendor side were the manufacturer of Masada's current computing equipment, the firms that bid on Masada's request and the Big Eight consulting firm. The first was interested in preserving and enhancing its existing stream of revenues from the customer. The bidders desired the profit potential in a new customer. HSS, in particular, saw in Masada the success of its plans for commercial viability. Big Eight expected to derive revenues from Masada, apparently had an axe to grind against HSS and possibly also believed it not in their interests to agree with the conclusions of the internal data-processing people. For Big Eight to have supported Rawls's preferences may have led Carling and Checking to feel dissatisfied with the value they had received for the fees they had paid the consultants, a situation no consultant would like.

In addition to the vendors, several other outside parties played important roles in Masada's drama: the members of the hospital association, who were displeased with Big Eight, and the administrators of HSS's existing clients, who saw their interests linked to the success of HSS in attracting Masada.

The differences in the interests of users and vendors are most sharply apparent in the process of evaluating the vendors' bids. Using the identical scoring scheme, Big Eight and Masada's data-processing people each ranked their favorite vendor higher than the others. However, neither of these parties was apparently considering the hospital's interests as viewed by its own top administrators because the latter eventually took over the decision process and chose a very different alternative.

The case of Masada illustrates the ways in which the differing interests of users and vendors and the differing interests among vendors can determine the technical features of an acquired computer system. User/vendor politics can also determine the design features of applications, whether operational, monitoring and control, planning and design, communication or interorganizational.

7

The System Infrastructure

INTRODUCTION

One aspect of the context of system building is the world of computing vendors. A second is the internal organization of people and resources devoted to computer-based systems—the infrastructure. *Infrastructure* literally means "the structure within." The term is customarily used by urban planners to denote the roads, bridges, utilities and other structures necessary for modern city life, but its meaning has been expanded to include the social, administrative and political mechanisms that support those tangible objects. When applied to systems in organizations, the term is used in its expanded sense, referring to both the tangible equipment, staff and applications and the intangible organization, methods and policies by which the organization maintains its ability to provide system services.

Organizations develop infrastructures around systems for a variety of reasons—for example, to provide a liaison with external vendors, to evaluate the competing offerings of multiple vendors, to avoid dependence on any one vendor, to provide a service in-house that cannot be obtained on the open market or to accomplish some combination of these objectives.

When a user and a vendor come together to negotiate, they must be sufficiently familiar with one another's worlds to recognize and solve all of the technical and organizational issues which arise in the sale of any complex industrial good. This is particularly true in the computing world, since computing technology is inherently unspecialized. . . . From the perspective of the user, there is considerable value in either confining oneself to a single integrated supplier on the one hand, or to develop a full-scale product evaluation organization internally. These alternative strategies of course, are merely tendencies, not absolutes; but they are clear-cut nonetheless, and the trend is toward making them more so. [Kling and Gerson, 1978: 37]

The development of computing capacity in-house reduces dependence on outside vendors and increases the ability of an organization to determine its computing destiny. However, the infrastructure exerts an influence on the shape of systems in organizations not unlike the influence of outside vendors as it increases in size and complexity. In fact, many organizations establish policies that require users to contract for all system services with an internal data-processing department, regardless of whether the services are ultimately provided in-house. This sets up a monopoly vendor inside the organization on which the organization may become just as dependent as it would on a single vendor or a service bureau. While an internal monopoly provider may be less likely than external vendors to act against the interests of users, it is nevertheless true that organizational subunits develop goals and objectives of their own that do not necessarily correspond to those of the organization as a whole or to those of other subgroups.

SYSTEM PROFESSIONALS

One of the most visible characteristics of the infrastructure is the variety of jobs, tasks and roles among those who provide computing services. Since the earliest days of computer use, the skills and knowledge required to make the computer work have been packaged in the form of computer specialists.

Originally, there was only one type of computer professional, the programmer. Programmers developed algorithms, wired plug boards or punched program cards and paper tape and operated the computer. Since then, the role of the programmer has differentiated into a large number of specialties.

One of the first steps in this differentiation was the separation of systems analysis from programming. Apparently, this innovation in programming structure was first tried on the SAGE (Semi-Automatic Ground Environment—part of an "early warning" radar network) project, during the Korean War. It was viewed as an attempt to monitor and evaluate programmer performance and to improve productivity because not enough trained programmers could be found. Creation of the systems analysis function separated "the conceptual tasks of designing a program from the more mechanical tasks of writing down the detailed instructions or 'code'" (Kraft, 1979: 145).

Another step was the separation of tasks relating to software development from tasks relating to the operation of equipment and production (i.e., running) software. This allowed factorylike methods of organization and supervision to be applied to the routine and predictable operational processes in computing, while engineeringlike project management techniques were adapted to the less structured software development processes.

A third step occurred within software development. This was the differentiation between systems software development and application software development. (See chapter 6 for the distinction between systems software and application software.) The former, called systems programming, was taken over almost entirely by external vendors. The systems programmers who work in computer-using organizations today primarily install and maintain rather than create systems software. The latter, called application programming, is performed by application programmers. These people, who work in software houses and computer-using firms, perform all phases of application development: They develop, maintain and/or enhance business-oriented software.

Job Categories

Since these early days, the differentiation of computer-related jobs has accelerated. Figure 7.1 lists the most common job titles in the areas of application software development, computer operations, technical support and management. Not all organizations have all of these categories, and some group them differently (e.g., maintenance programmers in computer operations, data base administrators and quality assurance in technical support, etc.).

Personality Characteristics

The system professional has long been observed to differ in temperament, outlook and motivational need from other members of the organization that use systems. Many people believe that these personality differences between professional designers and users can explain difficulties that arise during system building, just as design features that do not match designers'

FIGURE 7.1
Job Categories of System Professionals

Application Software Development	Technical Support	Computer Operations
Programmers	Systems programmers	Console operators
Analysts	Telecommunications	Secondary storage
Programmer/analysts	specialists	handlers
Analyst/programmers	Network analysts	Data preparation
Maintenance	Hardware capacity	personnel
programmers	planners	(keypunchers)
Data base	**Managerial Personnel**	Data entry clerks
administrators		Job control clerks
Data base	Vice-president,	Data librarians
programmers	director or manager of	Output handlers
User liaisons,	information systems	Scheduling
representatives or	Manager of software	
coordinators	development	
Business analysts	Manager of technical	
Operations	support	
researchers	Manager of operations	
Methods specialists	Standards and policies	
Quality assurance	Security	
Trainers	Manpower	
Information	development	
specialists		

intentions are said to explain user resistance and systems with detrimental impacts. A fair amount of research has demonstrated significant differences between designers and users, although the connection between these and system-building problems has not been empirically established.

Cougar and his colleagues (1979), for example, have administered the well-known Job Diagnostic Survey to several groups of data-processing personnel and managers and have compared the results to those obtained for general managers at various hierarchical levels. The Job Diagnostic Survey was designed on the basis of theories of human motivation. It measures:

· The extent to which jobs provide the motivating characteristics of skill variety, task identity, task significance, autonomy and feedback;

· People's job satisfaction;
· The extent to which the people tested experience the need for psychological growth and achievement.

Cougar et al. found significant differences between computer professionals and general managers on a number of dimensions. The most striking difference concerned the very low needs of system professionals and their managers for social interaction:

> Surprising or not, the results have important implications—in two areas: 1) the effectiveness of communication with others on the company management team, and 2) the effectiveness of communication within the computer department, with both superiors and subordinates. [1979: 53]

Thus, many of those whose role is to work with clients to determine needs and objectives for systems find more satisfaction in the technical aspects of their profession than in the social process by which technical specifications are determined.

The view of system professionals as almost asocial, when compared to the more gregarious managerial types, comes through again and again in scientific and humorous portraits of data-processing types. For example, one study reported in *Computerworld* (Stevens, 1980) prepared this profile of the data processing personality from psychological tests administered to 6,000 people:

> The DPer is more cool and impersonal than other personnel. He often prefers to be aloof when dealing with the others. . . .
>
> The DPer tends to be very perceptive. He is intellectually curious, . . . [But] he may respond poorly to supervision or leadership he feels is mediocre.
>
> The DPer tends to be more serious and reserved than his non-DP counterparts . . . , preferring to conceal his feelings. . . .
>
> The DPer tends to be more persevering, even dogmatic . . . tends to be moralistic. . . .
>
> The DPer is often more cautious and hesitant than his non-DP counterparts. . . .
>
> The DPer tends to be more pragmatic. . . . Concerned with immediate tasks and problems, he sometimes takes a narrow point of view. Short-range concerns are of interest, while long-range things are usually not. . . .
>
> The DPer is unpretentious. He is more straightforward . . . , even blunt. He may lack empathy in dealing with others. Choices made on social questions show that the DPer may feel socially awkward.

> The DPer is more anxious and uncertain. . . .
>
> DPers taking the test made more conservative choices than non-DPers. . . .
>
> The DPers observed were . . . joiners and were willing to be cooperative when group interests were at stake [more so than non-DPers]. [p. 25]

The following description of computer science undergraduates is not based on formal psychological tests but captures even more vividly the commonly held stereotype of computer people:

> In the middle of Stanford University there is a large concrete-and-glass building filled with computer terminals. When one enters this building through the glass doors, one steps into a different culture. Fifty people stare at terminal screens. Fifty faces connected to 50 boxes, connected to 50 sets of fingers that pound on 50 keyboards ultimately linked to a computer. If you go farther inside, you can discover the true addicts: the members of the Establishment. These are the people who spend their lives with computers and fellow 'hackers' [people who make constant small and unimportant modifications to computer programs for enjoyment rather than to improve the programs]. These are the members of a subculture so foreign to most outsiders that it not only walls itself off but is walled-off, in turn, by those who cannot understand it. The wall is built from both sides at once.
>
> These people deserve a description. In very few ways do they seem average. First, they are all bright, so bright, in fact, that they experienced social problems even before they became interested in computers. Second, they are self-contained. Their entire social existence usually centers around one another. Very, very few remain close to their families. Very, very few associate much with anyone who is not at least partially a member of the hacking group. While they do sometimes enjoy entertainment unrelated to their field, it is almost always with fellow hackers. Third, all aspects of their existence reinforce one another. They go to school in order to learn about computers, they work at jobs in programming and computer maintenance, and they lead their social lives with hackers. Academically, socially, and in the world of cash, computers are the focus of their existence. ["Hacker Papers," 1980: 63]

Computer professionals have been suspected of differing from users on personality characteristics other than needs for social interaction. One of these is cognitive style, the habitual ways in which people view the world and solve problems. Mason and Mitroff (1973), for instance, contend that most MIS designs have implicitly assumed that the users of systems will have the same cognitive style as those who design them. McKenney and Keen (1974) and Benbasat and Taylor (1978) among others, however, have found significant differences between users and designers on dimensions of cognitive style. Consequently, systems designed to support the thinking processes of those with one style would be quite awkward or inappropriate for people with a different style. Keen and Bronsema have summarized the argument as follows:

> 1. There are systematic differences among individuals in terms of perception, thinking, and judgment that significantly influence their choice of and response to information.
>
> 2. The difference between managers' [users'] and analysts' [system designers'] cognitive styles is a major explanation of difficulties in [system] implementation. [1981]

Other researchers have been less concerned with the problem-solving styles of computer professionals than with their assumptions about people. These assumptions may, they argued, be reflected in systems that exert excessive control or that monitor users' performance too closely. Hedberg and Mumford (1979) described a study of Swedish and British system designers. They found that British designers want the work of employees to be well defined, structured and stable with carefully set targets and close supervision. Swedish designers are more willing to believe that people are motivated by challenge, flexible jobs, few controls and self-discipline. For both groups, however, there were recognized differences between their theoretical or espoused models of man and those they used in practice to design systems. The espoused models viewed successful organizations in terms of good human relations while the operational models emphasized structure, control and close supervision:

> At the organizational level systems designers perceive themselves as having a rather limited role. Their contribution is to assist the organization to become more efficient through speeding up work procedures and providing information that is useful to management. . . . [They] have no conception of themselves as organizational designers using computer systems to produce new organizational forms. [Hedberg and Mumford, 1979: 48–49]

Taylor (1979) carried out a similar study of production engineers, system analysts and their managers and found similar results. Both groups used the principles of assembly line mass production to guide them when they designed jobs. Dagwell and Weber extended these studies to system designers in Australia with comparable results. They describe the process by which they theorize designers' models of people to influence system outcomes:

> Designers build up a conception of an "average" user in terms of certain attributes: the extent to which users need variety in their work. . . . [This] is the designer's model of man. To achieve an effective and efficient system

design, designers both consciously and subconsciously produce designs that take into account the perceived strengths and weaknesses of the system's intended users. For example, if the designer perceives that users are not responsible and cannot be given autonomy, it is likely that the resulting design will allow users little control over the nature and pace of their work. [In press]

In summary, then, evidence shows reasons to believe that system designers differ psychologically from users in ways that may affect the nature of systems built and, consequently, their affects on organizations. How the computing infrastructure is organized and managed may exacerbate these psychological differences by segregating users and designers and by applying different criteria of reward and evaluation to them.

Occupational Identity

Some of the difficulties in system building are attributed not to the personality characteristics of designers but to the computing occupations. The argument is that system professionals owe their primary allegiance to their profession rather than to the organizations in which they work. Therefore, they have many common interests with each other and few with their noncomputing colleagues.

Sociologists have observed that professionals who work in nonprofessional organizations, such as engineers, accountants and lawyers, often show at least as much loyalty to their professions as to the firms in which they work (Gouldner, 1954). Computer workers, with their trade associations and publications, may have a stronger professional orientation than organizational identification, in contrast to general managers who tend to identify chiefly with their firm. This orientation may lead computer professionals to advocate courses of action that managers view as not in the best interests of their firm.

Many examples of this process can be found. Although things are changing somewhat, software development workers have for a long time demonstrated distinct biases against software packages purchased from outside vendors. Because of the high cost and long time delays associated with in-house development versus package purchases, managers tend to view this as contrary to their interests. Computer professionals often prefer to build software in-house rather than buy it because, for example, "vendor software does not always follow the 'XYZ' method of construction and its documentation is not up to our standards." This tendency is so widespread that it has been named the NIH (not-invented-here) syndrome.

Another example is the tendency of computer professionals to acquire the latest hardware and software, even when clear benefits do not result from this choice. Here, professionals are motivated by concerns for their careers.

Career ladders are relatively short for computer professionals in organizations in which computing is not the major line of business. Because of this, career mobility between organizations becomes an important way to advance and to increase compensation. The interorganizational mobility of computer professionals depends largely upon the degree to which their skills remain up to date and transferable rather than narrowly specialized from experience in a technical environment that has become obsolete. Therefore, professionals will be motivated to bias equipment and software decisions in ways that do not always correspond with the interests of managers.

Thus, the career considerations of computer professionals can constrain and influence the outcomes of vendor and product selection, as in the case of Masada Hospital in chapter 6. Gerson (1978) has discussed these constraints at some length. He points out that, unlike the situation in the early days of computing history, computer operations managers and workers today have few opportunities for advancement. Application programmers may retrain themselves as systems programmers but have little overall mobility inside the firm. Systems analysts have the greatest, but still limited, opportunities both within and outside data processing.

Furthermore, career prospects and exposure to new skills depend quite heavily upon the size of the data-processing shop in which the professional works (Gerson, 1978). In larger shops, the division of labor is elaborate, offering workers the ability to exercise a very limited range of skills. In contrast, exposure to a variety of technologies may be greater than in a small shop, and explicit career paths (within the occupational class) may be present. In small shops, computer professionals may have considerably more opportunity to learn skills outside their occupational specialties, since these shops are usually informally organized. However, small shops offer limited career advancement possibilities and limited exposure to varied technologies. Consequently, one can expect the constraints on vendor and product evaluation posed by the computing infrastructure to be significantly different for large and small data-processing departments.

Another variation of this argument holds that problems in system building derive from the systematic efforts of managers to control, routinize and deskill the work of system designers (Kraft, 1977; Greenbaum, 1979). Managers divide system design activities into small tasks, reorganize these into repetitive jobs, and impose oppressive patterns of work organization on programmers and analysts in order to reduce dependence on system professionals and to cheapen their labor. One consequence is the alienation of software workers, who unionize or frequently move from one organization to the next. This encourages managers to continue removing skills from work so that workers can be more easily replaced. Ultimately, this degenerative process can result in professionals who are too highly specialized and too little attached to an organization to work in the best interests of the firm.

Attempts to control may work in both directions. The job hopping and turnover of data-processing professionals may be more than a response to managerial control and deskilling attempts; these behaviors help to exert power over the consumers of their services and to keep their salary levels artificially high by creating a perceived scarcity. Pettigrew (1972) presents a case in which members of an internal data-processing group were able to use their position inside the organization to improve their status and pay relative to data-processing professionals in the surrounding geographic areas.

Because of conditions like these that go beyond the boundaries of any one system-using firm, a basic conflict of interest exists between system designers and users. Users desire cheap software products of predictably high quality, whereas builders desire enhanced occupational status, prestige and pay through the performance of specialized, craftlike services that cannot be routinized, automated or obtained elsewhere. This conflict of interests may manifest itself in systems that do not meet the users' needs.

In this light, it is interesting to examine some trends in software production. Application development tools and end-user programming are two emerging software development technologies that appear to reduce the division of labor between system builders and system users. Application development tools are skeleton programs that can be completed and tailored by relatively unskilled software workers who do little more than fill in the blanks. End-user programming allows managers with very limited knowledge of programming to analyze data.

Application development tools clearly eliminate skills needed by programmers by replacing the writing of modules of code with simple responses to programmed choices. At the same time, this technology may allow a single software worker to develop whole systems or very large pieces of them rather than the unidentifiably small segments that typify traditional methods. Therefore, it simultaneously makes work simpler, but more important, reducing the degree of challenge in the job but raising the degree of autonomy of the professional who develops applications.

End-user programming allows users to reduce their dependence on software workers since they are able to do much of their own computing development work. At the same time, this technology may reduce users' level of dissatisfaction with software workers, leading to better communication on projects where the services of professionals are required.

THE INFRASTRUCTURE

The personality characteristics and occupational identity of system professionals may predispose them to interests and objectives different than those of system users. However, the interaction perspective suggests that the causes of system-building problems lie not so much in these uncontrollable factors but in the interaction of the system-building process with its context. One dimension of this context is the world of computer vendors discussed in chapter 6. A second dimension is the infrastructure, or the ways in which the many computer professionals are organized in system-using firms.

The infrastructure can be discussed at a number of levels: the organization of application development work, the organization of maintenance work and the overall structure or placement of application development vis-à-vis user departments, sometimes referred to as the centralization/decentralization issue. Each of these aspects of the infrastructure is discussed briefly in the following sections.

Organization of Application Development Work

The three most common arrangements for organizing application development are shown in figure 7.2. These are functional, project and matrix structures.

In the functional structure, analysts are located in one department and report to a manager of analysts, while the programmers are located in another department under their own manager. Someone within the first group is responsible for the early stages of the life cycle; at some point, project responsibility passes across departmental lines into the second group. Unfortunately for the success of this arrangement, many projects do not allow for such a clean distinction between the activities of each group: There may be long periods of time when analysts and programmers should work together to prevent miscommunication. Cooperation is difficult, however, when organizational lines must be crossed.

The project structure attempts to improve coordination by interposing a formal project leader between the manager of software development and the analysts and programmers who work on the project. Analysts and programmers report directly to the project leader either permanently or for the duration of the project. This eliminates the need to coordinate across departmental lines, but it reduces some of the benefits of having a single manager responsible for all analysts or programmers. These benefits may include more effective assignment of individuals to projects requiring special skills, more flexible scheduling of laborpower to activities, better career planning and skill assessment and increased training opportunities.

The matrix structure attempts to achieve the best of both forms of organization by combining them. Two types of managers are created, project leaders and functional managers (for analysis and programming). Analysts and programmers are assigned to projects for the duration of the project or a task.

FIGURE 7.2
Organization of Application Development Work

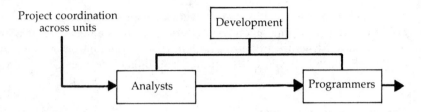

Project coordination
across units

(a) Functional Structure

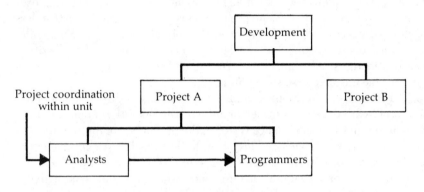

Project coordination
within unit

(b) Project Structure

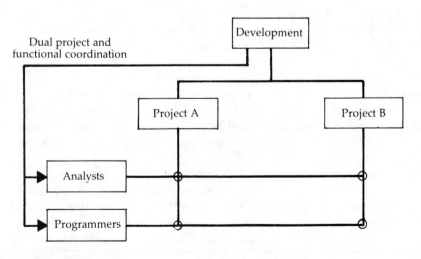

Dual project and
functional coordination

(c) Matrix Structure

They report simultaneously to two managers, one providing project coordination, the other providing functional direction and personnel administration. Because these structures require a high degree of communication among all parties, they are believed to be complex and difficult to manage.

Organization of Maintenance Work

The term maintenance is applied to the ongoing process of taking a so-called operational system—that is, one for which the initial analysis and programming is complete—and changing or enhancing it to meet evolving user requirements. Many analysts and programmers prefer the more creative, less constrained process of development to that of fixing someone else's code. Consequently, organizing for and staffing maintenance work are critical decisions in the system infrastructure. Several ways of approaching these decisions have evolved, as shown in figure 7.3.

In some organizations, maintenance analysts and programmers are located not with development personnel but with operational personnel, on the assumption that maintenance is part of the production of routine computing services. This arrangement may lead to poor quality systems since the developers do not have to face the consequences of their work.

In other organizations, maintenance work is a functional activity, separate from new development work but located within the analysis and programming unit. This maintenance group is responsible for fixing and enhancing all of the applications for the organization. Since these applications may employ many different languages and underlying technologies, it is often difficult to find people with a broad enough range of skills (as well as high enough tolerance for variety and time pressure) to handle the work load of this department.

A third way of organizing maintenance is to include it within the project development structure. As projects move along in their life cycles from development to maintenance, the work of the project team assigned to them changes, too. Programmers and analysts who prefer new development work or who desire the variety of different assignments often dislike this arrangement.

Personnel policies may soften or sharpen the effects of these structural arrangements. Some organizations, for example, allow people to express a choice for or against maintenance work rather than making arbitrary assignments. Others use maintenance as a training position for new analysts and programmers. Nevertheless, each arrangement has advantages and disadvantages in interaction with the life cycle and techniques of system building.

Figure 7.3
Organization of Maintenance Work

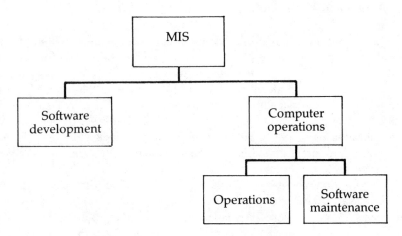

(a) Maintenance in Operations

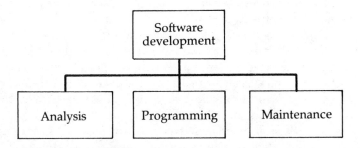

(b) Maintenance as a Function in Software Development

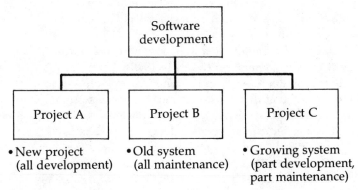

(c) Maintenance in Projects

Centralization/Decentralization

The discussion so far has emphasized structural arrangements within departments devoted to software development without examining the relationship between these units and user departments. This relationship is usually discussed under the heading of centralization/decentralization. Unfortunately, this choice of heading reflects the interests and concerns of those who manage rather than those who use computing. The great debate over the centralization or decentralization of computing has focused more on how to control the resources of people, machines and dollars than on how to control the impacts of computing.

Questions about how to organize the computing resource did not arise in the earliest days of computer use (see appendix B for a brief historical review). Computers were too small to handle all of the problems that could be delegated to them, so each department that could justify the expense got its own. It was only when expensive, large capacity computers appeared that serious questions of organization and control began to be raised, and then economics and technical considerations drove the decisions. If the economics of hardware favored centralization, as was the case in the 1960s, then both hardware operations and software development work were centralized.

The economics changed in the 1970s, however. Hardware began to arrive in smaller and cheaper packages, and the economics of software became a major concern. People began to realize that a decision to centralize or decentralize could be decomposed into two or three subdecisions: What to do about hardware, what to do about software development and what to do about everything else such as policies and standards—in a word, management (McFarlan et al., 1973). The prevailing wisdom was that computing management should always be centralized but that hardware and software development could be organized independently of each other. Therefore, in addition to the two pure types of hardware and software development—centralized and decentralized—a number of hybrid types could be identified where each dimension was organized differently.

Withington (1972) identified five hybrid organizational structures for computing that were common in the early 1970s. Large organizations with highly technical products—aerospace firms, for example—frequently centralized the operation of their scientific and engineering processing, leaving system development to user divisions. In large businesses with geographically dispersed divisions performing identical functions, system development was frequently centralized, while operations were decentralized. In large, dispersed organizations with similar but not identical divisions, Withington observed a tendency to centralize the acquisition of decentrally operated hardware and to develop applications common to all divisions centrally.

Smaller and less diversified organizations frequently adopted a centralized computer center and a centralized development group, augmented by smaller satellite centers and groups for unique local needs. In multinational conglomerates, Withington saw a tendency to centralize standards and policies for equipment acquisition and personnel training and to develop common systems for management reporting but to decentralize all operations and most system development. Figure 7.4 diagrams a few of the common arrangements.

In the late 1970s and early 1980s, the economics of computing began to shift again. The proliferation of personal computing, both standalone and tied into large-scale mainframe computing, has encouraged organizations to take a new look at their managerial policies and procedures and at the structure of

FIGURE 7.4
Centralization/Decentralization Alternatives

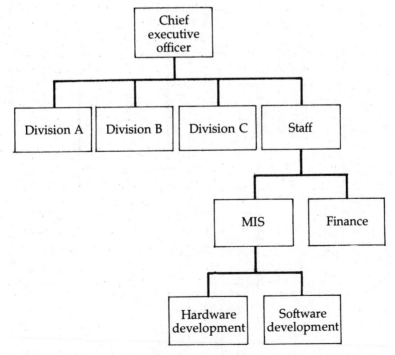

(a) Centralized Hardware Operations and Software Development

FIGURE 7.4
(*Cont.*)

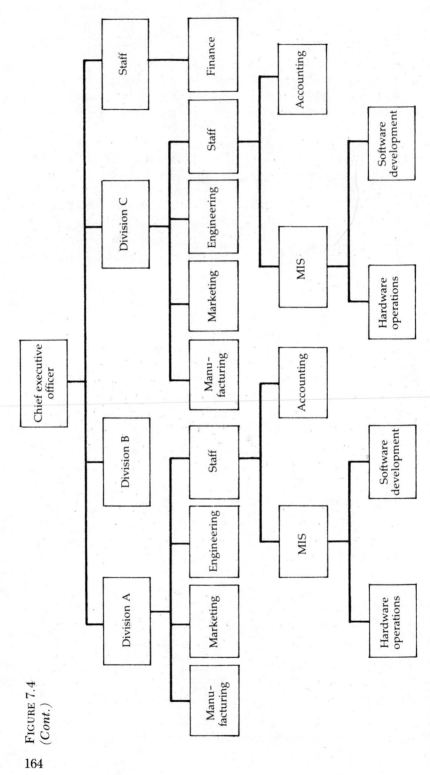

(b) Decentralized Hardware Operations and Software Development

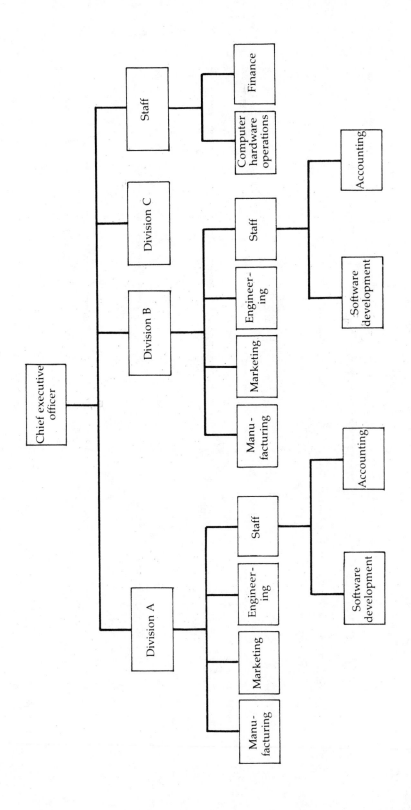

(c) Centralized Hardware Operations and Decentralized Software Development

FIGURE 7.4
(*Cont.*)

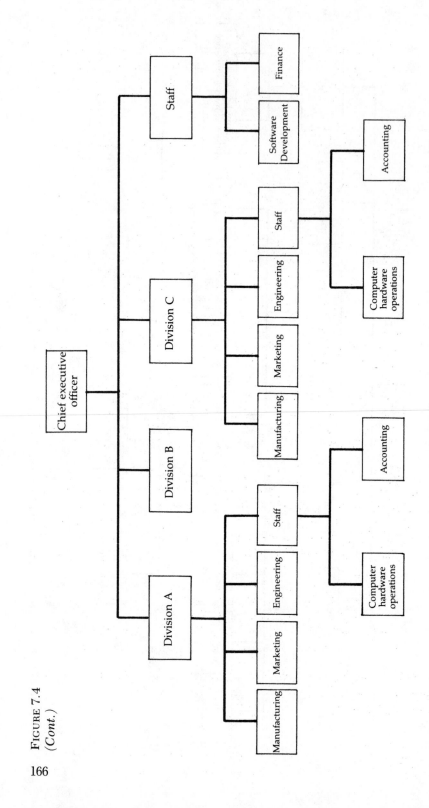

(d) Centralized Software Development and Decentralized Hardware Operations

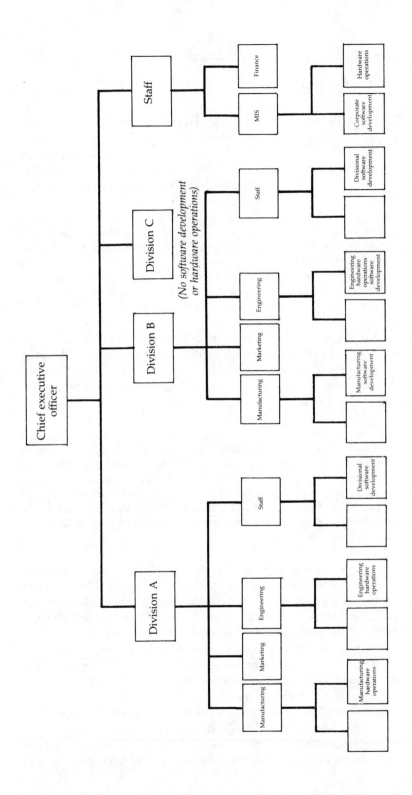

(e) Partially Centralized with Local Option

167

computing in their firms. We cannot describe the shape of the infrastructure
that will support the emerging technological environment of systems, but the
trend is away from the clear-cut structures of the 1970s.

For example, Rockart et al. (1977) have based their prescriptions on
the assumption that organizations should not make a monolithic centralization
or decentralization decision. Different subunits and application types might
benefit from separate treatment, as this quotation suggests:

> Even within a single organization . . . field offices might be geographically
> dispersed with decentralized management, while accounting and payroll are
> centralized at headquarters. . . . Hewlitt-Packard is an example of a multisys-
> tem organization. It has built a variety of coexisting systems—centralized,
> decentralized and distributed—including a 110-node worldwide telepro-
> cessing network. . . . computing to its present mix. [Shirey, 1980]

To accommodate differences in subunits and applications, Rockart et
al. (1977) have recommended that computing be organized functionally not by
the three activities of application development, computer operations and
system management but by what they call "logical application groups." A
logical application group is:

> [A] complete application system, such as order entry. In some cases it may,
> however, be an application subsystem (e.g., only order editing, credit
> checking, and inventory updating) or, rarely, a single application. The
> significant factor underlying the concept of the [logical application group] is
> that each is concerned with a logically separate process performed by the
> organization. As a result, although there usually is some transfer of informa-
> tion between [logical application groups], these transfers are minimal and
> well defined. [Rockart et al., 1977]

The relative independence among application groups allows a combined
decision about decentralizing both application development and hardware
operations to be made for each one without affecting the others. This approach
seems better suited to the technology of the 1980s than a straightforward
choice of one of the basic hybrid structures shown in figure 7.4.

Decisions about the centralization or decentralization of computing
are very important to system-using organizations. From the point of view of
those who manage the computing resource, there are advantages and disadvantages to each type of arrangement. Rockart and Leventer (1976) have
carefully enumerated these, a summary of which is presented in figure 7.5.

One factor contributing to the centralization/decentralization decision
is the physical placement of computing equipment because of the belief that
programmers must be located near the computers on which they work. This

FIGURE 7.5
Advantages and Disadvantages of Centralization/Decentralization

	Centralized	Decentralized
Driving Forces	Information interchange Control Government regulations	Economics: hardware, telecommunications Need for fast turnaround time Organizational/management philosophy Technology
Advantages	Consolidation Integration of function Common systems Best use of people given labor shortages	Lower communication costs Motivation
Disadvantages	Contention schedule priority Risk of failure Political problems Rigidity	No professional electronic data processing Idle resources Equipment incompatibility

SOURCE: Adapted from John F. Rockart and Steve Leventer, 1976.

belief made sense in the 1950s, when programmers often operated computing equipment, and in the 1960s and early 1970s, when most program development was conducted in a batch mode. Today, however, when many programmers work from terminals in an interactive mode, the co-location of programmers and computers is no more essential than the co-location of any other group of users and their computers. In fact, centralized development groups have been organized even in situations where the hardware is decentralized or distributed. The rationale for centralizing development in these cases centers on the need for a critical mass of technical people. Such a critical mass is said to foster skill specialization, to reduce personnel turnover and to eliminate duplicated effort.

The co-location of software development personnel and users seems more critical to the development of good applications than the co-location of development personnel and computers. When application developers cannot see the problems of their clients or interact with them frequently enough to learn their languages and ideologies, the systems they develop are unlikely to address business and organizational problems. Thus, whatever the reason for centralizing system development personnel, the arrangement works best when the organization is small, centralized and geographically located near the central development group. In the absence of these favorable circumstances, a centralized software development group may increase its responsiveness to users by subdividing development personnel and assigning them permanently to key client groups like divisions or functional groups. This strategy gives a centralized structure the appearance of decentralization.

Consequences of the Infrastructure

Computing managers usually favor centralization over other arrangements because it offers the best opportunities for efficient use of, and reasonable control over, computing resources. Most users of systems have very different interests, however. Normally, their organizational units consume only a fraction of the organization's total computing resource, and consequently, their concerns favor convenience of computing services delivery rather than efficiency of services provision. The ideal arrangement for the computer user is decentralization to the point where at least software development, and possibly also hardware operations, is provided directly under the user's responsibility. In spite of the problems associated with having to manage one's own computing services, this arrangement gives the user control over the timing of services and the effort expended on providing them. In this way, the user can avoid contending with a central provider or, worse yet, with other users who claim higher priorities for faster service.

> [The capabilities of computing] generate a corresponding and sometimes unexpected set of problems for many computer users. People who use computer systems for a variety of daily tasks must adjust to changes in computer systems, vie for adequate priority for their computing jobs, develop backup procedures when automated systems fail and periodically search for skilled programming staff. As a result, the very technology which was supposed to be an unobtrusive aid and time-saver can become very attention-demanding and a source of continual low-level conflicts. The "problem solving instrument" is capable of generating its own special problems. [Kling and Scacchi, 1979: 107]

Other things being equal, then, the managers of computing will prefer centralization because it gives them maximum control, although it also maximizes the hassles of mediating among contending users. They tend to resist decentralization, unless the problems become unmanageable, because this reduces their domain of influence. Users prefer decentralization, however, since that arrangement provides maximum control to them, at the expense of the difficulties of managing the computing resource.

Many organizations have adopted hybrid structures between pure centralization or decentralization precisely to increase user control (or perceptions of it) over the design features of systems without diluting professional control over resource expenditures and equipment purchase. Examples of hybrid structures include steering committees of users to negotiate priorities and policies and liaison positions within the users' organization to smooth transactions with the central group. As early as 1968, sociological research on the computer profession described the rationale for hybrid structures and lateral coordinating mechanisms:

> The head of a line division is of high status in the organization, but the chances are that he understands little about what the computer (and the data-processing staff) can and cannot do for him. . . . If the situation goes unchecked, it is likely that he will have to deal with a member of the data-processing staff who occupies a fairly low status in the organization—a programmer or systems analyst who has no supervisory responsibility but who is of high expertness because of his knowledge of the computer. Co-operative relationships are very difficult to maintain under these circumstances. . . . Disagreements, non-cooperation, and avoidance behavior thus characterize relationships between superior non-experts and subordinate experts; this is often called line-staff conflict. . . . It is hypothesized here that many organizations . . . adapt to this situation by altering their formal structure. In particular, the role of the consultant [user liaison] . . . is institutionalized. [Meyer, 1968: 258–259]

It is important to note, as Meyer does, that "one could interpret the difficult relationship between line managers and data-processing specialists as a human relations problem [or a problem of personality differences], but [it can also be considered a] result of a type of bureaucratic structure that compels high-status non-experts and low-status experts to co-operate with one another" (1968: 259). Conflicts and problems that can adversely affect the quality of systems in an organization can often be traced to the specific ways in which jobs are differentiated, organized into units and coordinated to manage critical interdependencies.

Consider the role of user liaison in the system development process. Sometimes this job is primarily administrative, a point of contact for coordinating requests for services to the computing group or an overseer of the unit's computing budget. Sometimes, however, the occupants of this role are assigned responsibility for performing several steps in the software development process. When this happens, miscommunications based on personality differences and occupational identities may be augmented by miscommunications that occur because the process of building systems must be coordinated across an organizational boundary separating users and an internal group for providing computing services.

An example helps to illustrate these points. The case of Telecomm describes the organization of the computing infrastructure in a real company, the rationale for organizing it in this manner and some of the trade-offs this structure entails. In addition, it demonstrates a few of the problems associated with slicing up the applications development life cycle into different jobs and locating these on opposite sides of an organizational boundary.

THE CASE OF TELECOMM

This case concerns the data-processing department (DSO) and one user department, marketing, at Telecomm Corporation (not its real name). Telecomm supplies communications services to business customers. The marketing department consists of three major groups: (1) a sales unit, (2) a unit that prepares the marketing strategy for two product lines and (3) a unit that supplies research and systems support to the strategy group.

The Marketing Systems Organization

The Marketing Information Systems (MIS) group is responsible for translating marketing's needs for systems into a form that could be interpreted by the programmers in DSO. The MIS group has several subunits (see figure 7.6): a long-range planning group and system manager groups organized around the major clusters of marketing systems.

Several system managers are located within the system manager groups; they are responsible for managing the life cycles of their systems and for performing some steps in the development process. Among these responsibilities are the following, taken from an internal memo describing the management and development of business and information systems at Telecomm:

> Provide overall Project Management of individual departmental projects to assure their completion in a cost effective and timely manner and to assure they serve the users' needs. . . .

> Provide the necessary resources and expertise to perform the system analysis, functional system design, personnel subsystem and other departmental functions required to successfully implement departmental development projects. . . .

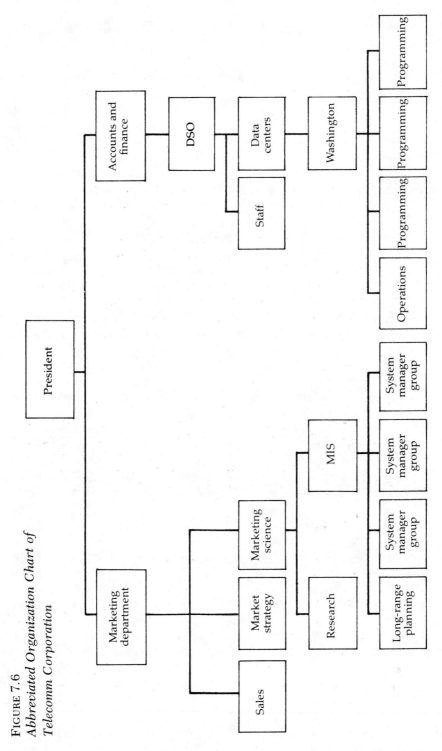

FIGURE 7.6
*Abbreviated Organization Chart of
Telecomm Corporation*

173

Provide overall System/Production Management of departmental business and information systems.

In short, the marketing system managers have to get and keep marketing's systems running. To do this, they must perform the initial phases of system development themselves, contract with DSO for the remainder of system development and for operation on DSO's computers, while managing the funding and approval process (described in a later section).

The Computer Professionals' Organization

DSO reports to Telecomm's vice-president of accounts and finance and, quite simply, centralizes three aspects of computing at Telecomm: hardware, software development and managerial control. DSO runs several large data centers, each consisting of several subunits. One of these is operations, consisting of the computers and those who operate them. The others are software development units, consisting of teams of programmers. Each software development unit is dedicated to a particular computerized system or group of systems.

The Washington data center develops and operates systems primarily for the marketing department, but not exclusively, and also runs some systems for other departments in Telecomm, for two reasons. First, some systems are technically interdependent with the marketing systems. Second, DSO managers wish to balance the work load among the data centers. Consequently, only two of the three programming units in the Washington data center correspond in scope to two of the three system manager groups in MIS.

The staff unit of DSO develops standards, embodied in system development practices, to assure the quality of systems developed. The most common type of standard involves documentation, which aims to provide the capacity to maintain (fix, change, enhance) the software over time, independent of the people who originally developed the system. Documentation, however, while solving one problem, creates others because most programmers do not like to do it and because good documentation takes time, which adds to the length of the system development process. Another function of DSO staff is to administer the project approval process.

The Approval Process

The approval process was designed to allocate resources among the projects competing for them and to serve as a quality check on the systems developed.

According to an internal memorandum, the approval process at Telecomm:

> [I]s a series of specified steps proceeding in a systematic manner from initial
> conception of an idea to verification that the project has met its objectives.
> Each step contains specific activities and objectives that must be satisfied
> before the next step begins. In sequence, the steps are as follows:
>
> 1. Data Center Assignment Request
> 2. Proposal Phase
> 3. Flexibility Phase
> 4. Definition Phase
> 5. Preliminary Design Phase
> 6. Detail Design Phase
> 7. Conversion Phase
> 8. Implementation Phase
> 9. Completion Report/Performance Review.

Thus, we see that the funding approval process is intertwined with every stage
of system development and that continued project funding is contingent upon
meeting the standards of procedure and documentation established by DSO
staff.

Very small projects (under $15,000) can be approved by DSO mem-
bers at the district manager level. Projects up to $50,000 can be approved by
the secretary of the systems steering committee, a division-level manager in
DSO. If the estimated costs of major development projects fall short of
$500,000, the initial proposal and subsequent phase reviews are submitted to
the systems steering review panel, composed of director-level managers from
each of the departments and DSO. This group meets monthly. Very large
projects (exceeding $500,000) are approved by the systems steering commit-
tee, composed of the departments' vice-presidents, which meets quarterly.

Few people at Telecomm question the value of the approval process as
a mechanism for allocating funds, but some decry the costs of the process in
time and money. Others question its efficacy as a quality assurance
mechanism:

> Systems are OK'd in phases. We only see the documentation for the current
> step, not for all that went before. This can allow people to change the scope of
> a project as it progresses, sometimes producing a duplication of resources. If
> people have done a lousy job of setting requirements, it has to be made up in
> later phases, or the designs are not maintainable or responsive to changes.
> The approval process should catch this sort of thing. In the six months I've
> been here, I've seen fifty projects reviewed, out of which forty-seven have
> been reports asking for funding. I'd like to see the approval process become
> part of a process to review the design document, because there are seven

phases of development behind the documentation of a request for funding. What gets seen in the approval meeting is a two page deal—with no real design information. We need instead a two-page structured walk-over. They shouldn't just approve the dollars. If this were a request for money to build a car, they'd demand to see the prototype, or at least the sketches. They'd say, "I want to look at it." Maybe we could change the approval process so that they review prototypes, look at the screen designs which show what the user does now and will do in the future. There are two aspects to funding: 1) reviewing the product, and 2) reviewing how it'll be produced. They think they're doing it all now, but they're not. [Personal interview]

The Interaction of the System-Building Life Cycle and Structure

Altogether, then, the six major organizational subunits involved in systems development are (1) clients in the department with needs for data and systems, (2) MIS system managers groups, (3) MIS long-range planning, (4) DSO support staff, (5) DSO programmers and (6) DSO data-processing centers.* The first three groups report through the marketing department, the last three through accounts and finance. Each of the groups is structured on a different organizing principle (see figure 7.7).

> We in MIS are, unfortunately, organized by systems. We should be organized by client interface. If the organization changes, as it does and has done, the system is not responsive to the organization. Systems should support the work process, not the organization.

When the approval process is mapped onto the organizational structure created to support it, it is clear that, in the course of development, projects often pass from the domain of one group into that of another.

> The process of developing systems is a pretty lousy process. It's a mess, almost doomed to failure. Well, look at it: here we set requirements, DSO does the coding and DPC [Data Processing Center] does the operations. So any job must go from me to SMG to DSO to DPC. The chain is really involved at the worker bee level, and, when you take into account the management mating dance, . . . [Personal interview]

When projects pass from system manager groups to programming, they cross a gulf that is bridged only at the very top of the organization.

---

*Adding the steering committee and the review panel to the list would bring the number of key players to eight. An internal Telecomm memo sets the number of persons involved in the approval process at fourteen.

FIGURE 7.7

Key Actors in Application Development at Telecomm Corporation

Users (Marketing Department)			Professionals (DSO)		
Clients	System Managers Groups	MIS	DSO Staff	Programming	Data-Processing Center Computer Operations
Organized by product and function	Organized by major system groupings as they relate to client	Support group, system manager group, interfaces with DSO staff	Support group for programming and operations, administer approval process	Organized by major system groupings as they interrelate technically	Organized by hardware and by shift

> When I got into this organization, the first thing I did was to draw the organization chart and ask: where do the users and data processing meet? The answer is the President of Telecomm. [Personal interview]

When a project crosses the boundary from marketing to DSO, a handoff is said to take place. Handoffs are believed to create two kinds of problems for the organization. From the perspective of marketing, the endless checking in the approval process slows projects down to the point where clients and system managers become sorely frustrated. System manager groups blame their lack of organizational control over DSO for this state of affairs:

> One other problem is that once we system managers have put together a package of specs, we have no input or control over time and cost. We have no input in DSO in these matters. DSO is always late. Why? They've got no responsibility to me. I don't control them. [Personal interview]

They point out, further, that not only is DSO autonomous but also it is a monopoly provider of its services.

Years ago there was rampant proliferation of computing at Telecomm. Telecomm decided to centralize all software development into DSO; *nobody else is allowed to program*. DSO developed a very powerful organization, because frankly, everyone else needed them. They do it, even if it is cheaper and more efficient to do it ourselves. We do the modeling here; we could do the whole thing, coding and all. If I could think of a way to do it without getting into trouble, I'd try to avoid the Steering Committee and DSO too. [Personal interview]

System manager groups, however, attribute much of the so-called time problem to their own clients, who like to see systems built yesterday:

It's frustrating to develop user requirements because the whole process is shorter than it should be. You develop an implementation schedule and they shrink it on you. They eliminate parts of phases. Users like to see an assignment done in six months. Lots of our systems should take more like the estimated 1-1/2 years. But they say they need it yesterday. So, we miss things; they suffer. We started off on a project, but it remained in limbo for four months. Then they came to us and said: "We'll put the system up if and only if it's up by August." We said: "That's crazy. Better to stay with what we have." But they insisted. We said: "Aye-aye, sir." I've become really hard-nosed with users. They tell me what they want and I propose the ultimate system. If they say the time frame is no good, I say: "Well, what do you want to eliminate?" I don't try to deliver everything in an unreasonable amount of time. [Personal interview]

Further, these groups believe that users sometimes change their needs or their minds between phases of the process:

Marketing and DSO haven't got off the ground in their relationships with each other. Marketing is bucking DSO for being too slow. DSO says we don't meet their standards. They say our specs are incomplete and that we are the worst offenders. But, what can we (MIS) do? The users keep changing their minds. [Personal interview]

The Impacts of System Building at Telecomm

Consequently, system managers are caught between two groups with very different orientations and reward systems. On the one hand, system managers are organizationally and geographically co-located with their client groups, whose orientation is to develop systems as quickly (and perhaps also as inexpensively) as possible. On the other hand, many system managers identify psychologically with their fellow data-processing specialists in DSO, whose orientation is to produce high quality, well-documented and easily maintainable systems. Similarly, both clients and DSO perceive system managers as belonging to the other group (DSO and clients, respectively).

> The system manager is in an untenable role. He's not really in the end user department and shouldn't be unless he has the resources to produce a system. This may be a radical view. We're considered to be from the end user department, but we're not really. How responsive could we be under these circumstances? [Personal interview]

From the point of view of DSO, the time pressures felt by clients and passed on to system managers lead the system managers to rush through the most critical phase of project development: analysis. The object of this phase is to produce specifications used by programmers to produce the final system. If the specifications are incomplete or inaccurately reflect users' needs, the system will be perceived as bad and, thus, either scrapped or reworked.

> The process of developing systems here is like a game of "telephone": the project gets passed from person to person, group to group. It's no wonder things get dropped or forgotten. There are two critical interfaces: system manager groups with the user, and system manager groups with data processing. [Personal interview]

> The system managers are analysts, designers and human factors engineers. If their resources are limited or if the project is complicated, they can't do an adequate job in all of these roles, in addition to project management. Usually the systems analysis function doesn't get done. All the due dates are set for design, so that gets done. Then DSO ends up re-doing the front-end piece. This is why projects are so expensive. They drop their project management role; they may not recognize the need for it. Also, the approval process does take time. [Personal interview]

> System managers like to scribble a few words on a blue line [pad] and call it specs. [Personal interview]

> The specs have *got* to be good. We *never* have time to do them twice. [Personal interview]

In embarrassed tones, DSO talked of one system that was programmed three times as a result of just these problems:

> An outstanding example of a system failure in marketing was a forecasting system. Two years and $900,000 later, it was decided that the project should be discontinued. Why? The organization structure had changed and the system no longer supported the business process. If the system had been

completed just six months earlier, we still would have got some benefit out of it. [Personal interview]

Many people at Telecomm believe that a structural change will solve problems with the system development process, but they differ in their opinions about what the change should be. DSO, naturally, wants to reorganize so that the system managers fall within its sphere of influence. The marketing department wants to have organizational control over the programmers. This way, they believe, they will get more responsive service. However, for political reasons, too involved to be discussed here, changes of the kind desired by either group are not likely to occur. In the meantime, DSO continues to produce systems that are late and ill attuned to the needs of the people in the marketing department.

CONCLUSION

As companies increase their investment in computing, they often build up infrastructures, if not to provide services efficiently and effectively, then to protect against dependence on, and exploitation by, the inevitable external vendors. At the same time that it reduces dependence on external entities, however, the growth of a system infrastructure creates a group whose interests differ from those of the users. The process and techniques of system building interact with the organizational context of the system infrastructure and frequently produce systems that reflect neither the best intentions of users nor the best techniques of system professionals. The resulting design features may, in turn, interact with the context of system use, producing negative organizational impacts or systems that users resist or ignore.

The users' view of the computing infrastructure is humorously captured in these words of an anonymous software consultant, cited in a newspaper article.

> "The greatest thing that could happen," he says, "would be to cut their data processing budgets 75 percent. Then they'd have to discover that there are ways they could do the same job, maybe a better job, on what's left!". . .
>
> At most firms, computer functions are concentrated in one spot under one control—"the witchdoctors of the data systems department" the [consultant] calls them. This department, he adds, is the biggest roadblock to data processing efficiency. . . .
>
> Because they have all the specialized knowledge, the data managers are often left largely unmanaged [by corporate line managers] themselves, and waste and error can flourish unheeded. [Blundell, 1981]

The humor in this quotation masks a situation that is even less funny than negative impacts and system failure. A major consequence of the barriers erected by the infrastructure between users and system professionals is the apathy of users in preventing or redressing undesirable outcomes. Many users fail to take charge of systems in their organizations because another group of people has been assigned the responsibility for it.

I hope this book has demonstrated that something must be done to improve the contribution of systems to organizations. The task of chapters 6 and 7 has been to show why vendors and computer professionals cannot be relied upon entirely to do this. The purpose of chapter 8 is to examine, from the interaction perspective, what can be done about systems in organizations and the role of the manager or system user in doing it.

System Design and Implementation

INTRODUCTION

Chapters 1 through 7 should leave no doubt on one point: Systems can go wrong in many ways. Designers may seek to achieve political objectives through systems, and users may resist them or experience negative impacts from using them as they were designed. Also, the designers' intentions might have resulted in an effective system if its intended features had not been altered in the process of design. This alteration might be accidental, something that fell through the cracks during a handoff in system design, or it might be the result of political processes involving users, vendors or system professionals.

Given the diversity of parties and interests involved in a major design effort, it almost seems a miracle when a system turns out right from everyone's point of view. The normal workings of the traditional system design process in conjunction with the computing world and the infrastructure are too complex and fraught with conflict to produce successful systems routinely. That successful systems occur as often as they do is a tribute to the many hardworking managers, users and system professionals who take on more responsibility than they have designated authority.

The organizational perspective on systems is not pessimistic; it does not assume that nothing can be done about bad results. However, it does imply that reliance on the way things are supposed to work will not lead to success. Managers who would build successful systems can increase their chances by applying the insights of the organizational perspective before they turn their requests for assistance over to a vendor or a system designer. Applying this perspective entails three steps, none of which requires much time or energy, only a little thought:

1. Define the organizational context,
2. Constrain the design features of the solution,
3. Specify the design process and the roles of participants in it.

DEFINE THE CONTEXT

From an organizational angle, the first step in effective system design and implementation is to define the context of the proposed system. This means identifying, describing and understanding all relevant features of the setting in which the system would be used. Obviously, the key word here is *relevant*. A thorough study of the entire organization prior to system analysis would clearly be more trouble than it is worth—nothing would ever get done. However, failure to understand the context of a system could easily lead to the selection of an inappropriate solution or inappropriate design features of an appropriate solution. The trick is to define the context broadly enough to include the factors that will influence effects on the organization and system success or failure without performing analysis for its own sake.

Why Bother?

The major reasons for defining the context of a proposed system is to determine if a system, any system, has the potential to effect the desired outcomes. The interaction perspective suggests that systems will be resisted when they differ too sharply from the features of their contexts—that is, structure, politics, culture and so forth. Therefore, if the objectives for a system require design features that are incompatible with the context, the system is not likely to succeed in achieving its objectives, not at least without concurrent changes in contextual features. In short, systems alone, unreinforced by changes in structure, job design and compensation, are not likely to effect radical changes in organizational functioning. Careful definition of the context helps to determine the riskiness, or likelihood of failure, of a proposed system.

Even when a system appears to be the appropriate intervention for improving performance, certain design features may interact with contextual features to produce undesirable impacts. Defining the organizational context identifies those features that the system must have or must avoid to achieve the desired results.

The example of insurance benefits processing at Grant Electronics illustrates the problems of improperly defining the context of a proposed system.* The manager of the insurance benefits department (IBD) at Grant Electronics has wanted to automate its largely manual process for some time. Grant, however, has been growing rapidly and has chosen to spend its resources on customer support rather than internal development. Recently, the department manager has been given the resources for system development. She reviewed the problems and opportunities in her department, and on the basis of this review, she defined the features of the system that would be built.

The IBD enrolls Grant's 16,000 domestic employees in the health, dental, life, accidental death, travel accident and short-term disability insurance programs offered by major insurance carriers. The department also initiates payroll deductions for long-term disability insurance, dependent coverage or increased coverage. Finally, IBD monitors the carriers' processing of employee claims and processes claims for long-term disability coverage.

Sixteen people work in the IBD. One person handles enrollments to and cancellations from the insurance programs (about 350 and 250 per month respectively for this growing company in a volatile industry). Another person handles any subsequent changes (about 200 per month) in insurance coverage (as when an employee desires coverage for a new child). A supervisor handles exceptions and inquiries. The claims-monitoring unit, consisting of a supervisor and four claims administrators (with two different job titles), monitors the status of claims. There are three disability administrators. In addition, the department has a manager, an assistant manager, two secretaries and a benefits specialist.

Although the enrollment group consists of only three people, the IBD manager believes this operation has the highest priority for automation. It consists of repetitive activities with minor variations or exceptions. It is labor and paper intensive; mistakes cost money and upset employees. The operation appears to be totally contained within the boundaries of the IBD, ensuring easy implementation of change. In addition, the enrollment group is currently about at the limits of their processing capacity, and Grant is expected to continue to grow rapidly. Finally, the second most critical

---

*The data for this case were collected by Carol Darling, Genevieve Tchang and Mona Matsumoto as a class assignment for a course at the Sloan School of Management at MIT, spring 1982.

operation, claims monitoring, is scheduled to receive terminals accessing the insurance carriers' data bases in the near future; this will greatly assist them in answering employee inquiries and thus reduce this as a priority for automation.

In short, the department manager of IBD has given the system analysts assigned to her the charter of developing a system to support the activities of the three enrollment handlers. In so doing, she has precluded changes in structure and job design that might benefit the organization far more than the proposed system.

For one thing, this definition of the context of the IBD system does not address the job satisfaction of the employees in the department. A solution in which, for example, each person in the IBD performed enrollments, changes, claims processing and monitoring for a subset of Grant's 16,000 employees might increase the challenge, importance and variety of the job, thus improving employee motivation. At the same time, it would eliminate handoffs among enrollment, changes and claims that might be a major source of the errors that so upset the department's constituency. Solutions like this are precluded by limiting the definition of the system's context to the three enrollment workers.

Second, this definition of the context does not address critical interactions between the IBD and another department, payroll. The importance of these interactions becomes clear from a brief review of the detailed procedures performed by the enrollment administrators.

On the first day of work, new employees fill out a benefits enrollment card. This is one of three main inputs to the enrollment process. The others are a salary printout from the payroll system (the enrollment administrator checks to see that the employee's name is there) and a new-hire printout from the payroll system (a cross check and a source of employee address information for mailing). From these inputs, the enrollment administrators verify and complete the data on the enrollment card, assign a grouping according to carrier definitions, compute deductions and complete a data entry form for the payroll system, copy the enrollment card and distribute the copies to a large manual card file in the office, to the carriers, to the employee. Similar procedures take place for terminations and changes.

From this brief description, it is obvious that insurance benefits procedures are highly interdependent with payroll processing. However, payroll does not fall within the boundaries of IBD. An organization chart (figure 8.1) clarifies the relationships. The IBD is part of the compensation division, which also determines payment plans and benefits packages. The payroll system is administered from a department in a different division (personnel systems). And the computer professionals that help to develop automation for IBD are located in a third division of the human resources arm

186

FIGURE 8.1
Grant Electronics Partial Organization Chart

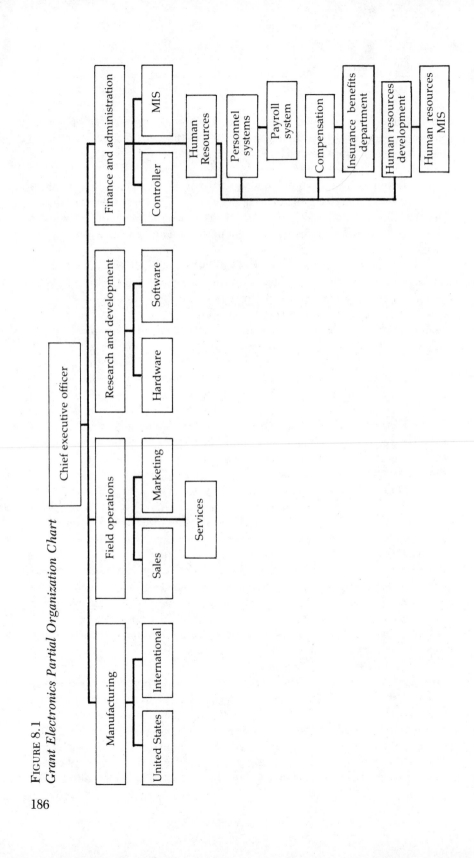

of finance and administration (the human resources development division), which is separate from the main MIS department for Grant's internal operations. Therefore, insurance benefits processing, far from being contained entirely within the IBD, is spread across several divisions, not just departments.

By limiting the context of the IBD system to processes falling entirely within IBD, the department manager precludes the development of a single system that would support both payroll and insurance benefits processing. Ironically, a system with the potential to support both payroll and insurance benefits had already been installed in the personnel systems division at Grant. (Even more ironic is the fact that this system had been developed by software engineers in Grant's research and development department as a product for sale to customers of Grant's computing systems.) This Human Resources Management System (HRMS) had provisions for insurance benefits processing, but when the system was installed in personnel systems in 1975, the insurance benefits features were never implemented. Because of the way in which the IBD manager defined the charter of the MIS people working for her, the option of activating this subsystem of HRMS was not considered.

This definition of the context of the IBD system also fails to address critical interactions between the human resources division, of which IBD is a part, and the rest of the corporation. Consider some additional data. In 1974, Grant Electronics had sales of $60 million and one person in the IBD (today she is its manager). In 1979, Grant had sales of about $300 million and about 6,000 employees, with 35 in human resources and 12 in insurance benefits. In 1981, Grant had sales of about $1 billion and 16,000 domestic employees, with 300 in human resources and 16 in IBD. Grant anticipates sales of $5 billion by 1990. Simple arithmetic suggests staggering increases in the size of human resources and the IBD.

At some point on its rapid growth curve, Grant will probably begin to consider divisionalizing and decentralizing its currently centralized, functional structure (see figure 8.1). At that point (if not before, given the projected size increases), it makes sense to question whether or not personnel processing (including both payroll and insurance benefits) should be decentralized to the level of the to-be-created business units.

This move has enormous ramifications for the design of the HRMS and IBD processing system. It is not currently within the state of the computing art to design a single system that can support both centralized processing for 16,000 employees and decentralized or distributed processing for three to five times this many employees. Therefore, a choice has to be made. If the decision is made today without considering the future, the design features of the system may ultimately dictate the design of the organization.

An expanded definition of the IBD system context reveals that the payroll-plus-benefits activity in Grant Electronics is not as centralized today as the original description made it appear. The original source of data about insurance benefits is the employee. Each of the manufacturing, field operations, research and development and corporate administration units has its own personnel function, quite independent of the corporate human resources unit depicted in figure 8.1. Currently, these line personnel specialists do little with the corporate HRMS other than distribute forms to employees, collect them, forward them to the corporate group and act as a point of reference for employee inquiries. These people are located in the ideal position, vis-à-vis employees, to be able to prevent and correct errors in data and to respond to employee inquiries—in short, to be the access point to a decentralized or distributed HRMS system.

Why Can't the System Analyst Do It?

System analysts have long recognized both the importance and the difficulty of defining the boundaries around systems. For example, in their book on system analysis, Gane and Sarson state:

> From the information in the initial study, and from the information gained in defining the user community, the analyst can now produce a draft logical data flow diagram of the current system. But what exactly is "the current system"? The user's concerns often are to do with business results in a specific area, say, purchasing. As the analyst investigates he may find several interlocking systems are involved, some manual, some automated. In these cases, it can be a problem to *define the system boundary,* deciding which functions are to be considered as part of the system study, and which functions are not . . . . Some organizations require their analysts to draw data flow diagrams of each system that interfaces with the area under study, especially the clerical systems. This requires additional time and effort, but has the benefit of showing up duplicated and redundant functions, giving the analyst greater confidence that he is drawing the system boundary in the right place. [1977: 246, emphasis in the original]

Why, then, should system analysts not be the persons to define the organizational context of a system? Why should managers identify the factors that must be considered in deciding whether to build a system and what system to build? The answer is that defining the context of a system often involves redrawing the boundaries around organizational units. Careful contextual analysis according to the interaction perspective frequently reveals the potential of structural changes to improve the functioning of the organization.

However, redefining organizational boundaries is a process that creates winners and losers among those who currently manage the units. Because of this political issue, redefining boundaries requires, at minimum, the strong commitment of the boss common to all affected departments. In many cases, MIS people do not even have access to managers at a high enough level in the organization. Further, they do not have the professional legitimacy to make suggestions about organizational design. In the early days of computing, system builders had a tendency to mandate structural changes by the way they designed their systems. This created so much resentment among managers that system professionals have almost entirely backed away from advocating structural change.

Consequently, system professionals advocate a design principle that states that systems must be independent of the organizational structures in which they are used (see e.g., Gane and Sarson, 1977). This design principle is quite defensible on the practical grounds that it is too expensive and time consuming to reprogram systems for small changes in structure. It has two unfortunate side effects, however. First, it prevents system analysts from identifying and recommending needed changes. Second, it encourages managers and system builders to try to use systems to solve organizational problems. An example of the latter type of thinking (in which the system professional was the good guy) is given by Turner:

> A designer was asked to build an integrated payroll personnel system for a firm. Including both functions in one system makes good sense since about 60% of the data elements are duplicated. However, in this situation the payroll department reported to one VP while the personnel office reported to another. The two departments were located in different sections of the country and they had a long history of interdepartmental conflict. The designer pointed out that management was asking the proposed new system to accomplish . . . an administrative action—the combination of the two departments. He suggested that the administrative change be made prior to the design of the system rather than burdening the system with the resolution of conflict, thereby reducing the probability of successful implementation. [1981b]

How to Do It?

Clearly, then, system professionals are ill suited to define the context of systems because of their training and their technology. But Turner's example and the case of Grant Electronics show that many managers also lack the ability to define the context of systems. Unfortunately, there is no foolproof recipe, but the organizational perspective in this book offers some guidelines.

First, do not use the boundaries of an organizational unit to define the context of a system. Do use a statement of the whole task or process. In developing a methodology for analyzing offices in preparation for office automation, Sirbu et al. had to wrestle with the issue of where to start, which easily translates into a question of contextual boundaries.

> Should one begin at the top of an organization and work down, or start at the bottom and work up? For a number of reasons . . . the starting point of most analysis projects will be a particular office or department that has requested the services of the office automation group. It is important, however, not to limit the scope of the study to the particular office in which it begins. [1981]

The authors recommend that the analysis focus on how business functions are implemented. They define functions as aggregates of activities that manage and maintain some resource that relates to the business goals of the larger organization. Two examples show what a function is and is not and why the boundaries of a system analysis cannot be drawn contiguously with the boundaries of an organizational unit:

> MIT's Office of Sponsored Research (OSP) carries out a single major function from beginning to end—managing research programs funded by means of a grant or contract from an outside sponsor. Here, the resources are "sponsored research programs." Several procedures are involved in managing sponsored research programs. The initiating procedure for a sponsored research program consists of the review and approval of research proposals from within the Institute and the negotiation of the terms of grants or contracts based on these proposals. . . . Once a research program has been successfully initiated, the OSP is responsible for administering the terms of the grant or contract that funds it, disbursing funds, approving travel and purchase requests, and monitoring expenditures. . . . On the expiration date of the grant or contract, the OSP must terminate the research program by closing out and auditing its account, and preparing a final report to the sponsor. Archiving the records of the research program is the final step of the terminating procedure.

> A university Admissions Office provides a contrasting example, one that demonstrates the value of the functional model [and the danger of using organizational boundaries for system analysis boundaries]. The mission of the Admissions Office is to review applications and admit or reject applicants. But this responsibility does not comprise an entire function—an applicant is not a resource of concern to the goals of the university, students are. An applicant is only a potential resource. Nowhere in the Admissions Office is there to be found a single procedure that is concerned with what to do with an applicant once he is admitted as a student, or how to deal with his records once he leaves the university. The overall function is that of managing the student body; the Admissions Office carries out only the initiating stage of

this function, that of admitting students into the system. If the integral connection of the procedures carried out in the Admissions Office to those performed in the Registrar's Office—which manages students once they are registered—[and the Alumni Office—which manages their records once they leave the university] is not recognized when an automation system is designed, then a major area of potential benefit will have been lost. [1981]

Sirbu et al. would argue, then, that the boundaries around a system's context be defined largely enough to include all tasks and procedures that initiate a system or follow from it. This is another way of saying that the boundaries around the system should include the whole process of which the system may be a small part. Taylor discusses this criterion in his description of a system analysis and design effort in a government agency:

> The government office . . . is one of six such offices distributed regionally throughout the United States, and which form part of a large centralized computer-based system of authorizing, processing and paying federal insurance claims. All the offices in the system had experienced problems of high work backlog, high employee absenteeism and turnover . . . and in shortage of trained staff.
>
> The analysis was undertaken in order to determine the causes of that poor technical performance and of the poor employee response to the work. We proposed that the boundaries of the entire structure of the San Francisco Program Center Office [including managers, supervisors, and nonsupervisory technicians and clerks], plus the computer and its staff in Washington, D.C. be used as the starting point for the socio-technical system. [The reason for this choice of boundaries is that] that system contained all the functional elements necessary for the primary mission of the office we studied, as an open socio-technical system [see discussion of autonomous work groups under Mumford's ETHICS method in chapter 5]. That system is in commerce with its environment of clients, lawmakers, and other parts of the agency. It takes in all raw materials (claims, existing records, and special information), utilizes a conversion process (programs and procedures), and exports the resulting products (payment orders) and services (updated records). [1981]

Underlying the whole-task criterion for defining a system's context is an organizational design principle, clearly articulated by Cherns (1976). The boundary location principle states that departmental boundaries should be drawn in a way that entire processes can be completed within the department, for two reasons: to avoid excessive costs of coordination across departments and to encourage employees to take responsibility for their work. Cherns acknowledges that "departmental boundaries have to be drawn somewhere" and that "all . . . criteria are pragmatic and defensible up to a point" (1976: 788). Grouping whole processes within single departments, however, has consistently produced better functioning than more typical forms of organization (like one that distinguishes between payroll and insurance benefits processing).

Second, do consider interactions that transcend the boundaries of the organization. Many systems involve transactions between an organization and its environment—e.g., customers, labor markets and suppliers.

Third, do recognize that any definition of a system's context is arbitrary and will need to be revised as new information becomes available. In their book on the sociotechnical systems approach, Cummings and Srivastva discuss three criteria normally used to set boundaries around departments: territory, time and technology. They show that these criteria can produce different boundary definitions:

> Let us assume that we are asked to examine a sales unit of a large manufacturer of heavy-duty machinery. . . . We make a field visit to the administrative sales office and ask a few questions of the sales personnel—supervisors, salesmen, secretaries, and filing clerks. Upon first entering the sales unit, we observe the following: (1) The [people] all work a standard 8:00 A.M. to 4:30 P.M. work day except for a group of filing clerks who file invoices on an evening shift for subsequent keypunching. (2) The [people] and some of the technological components—desks, files, typewriters, phones, desk calculators—all exist within a physical space bounded by two permanent walls and two temporary partitions. (3) The social and technological components all serve to operate a primary task of selling heavy-duty machinery. So far it looks as if we might have a well-bounded work system, with the possible exception of a group of second-shift, filing clerks.
>
> We proceed to interview some of the personnel. Our interviews reveal that: (1) The sales force is divided into four national regions and the salesmen spend about 70 percent of their time in their respective regions. (2) Supervision consists of a department manager and four regional supervisors. (3) The salesmen and supervisors are all male with college degrees, while the secretaries and clerks are predominantly female with high school diplomas. (4) The salesmen view their jobs in terms of making face-to-face sales, and the secretaries see their work roles as providing necessary clerical services to the salesmen. (5) The filing clerks perceive their jobs as similar to the secretaries with the exception of the evening shift clerks who view their roles as preparing materials for the computer services department. (6) The salesmen spend more time relating to regional customers than they do relating to members of the sales unit, and the secretaries and day-shift clerks relate to each other more than anyone else, with the possible exception of daily phone calls from the salesmen in the field. (7) The night-shift clerks relate to each other and to the keypunching operators of the computer services depart-

ment. With this additional data, the work unit is not as clearly defined as we had originally envisioned.

Although the boundaries are arbitrary, the above information allows us to explore some alternatives:

1. Define the entire department as one work unit.
2. Define the entire department with the exception of the night-shift, filing clerks as one work system.
3. Define the social and technological components for each region as a system, thus generating four work units.
4. Define the salesmen and supervisors as one unit and the secretaries and clerks as another unit.
5. Define the entire department and keypunching part of the computer services department as one system.
6. Include important customers in the definition of each regional work unit.

In working through this hypothetical example, we can begin to see that the issue of where to place a work system's boundaries is far from elementary. Although our criteria—time, territory, technology, and sociological and psychological attributes—can help to differentiate a socio-technical system, the result is still ultimately arbitrary. With this in mind, we initially bound a work system and continue to raise the issue of changing this definition in light of new information. [1977: 122–123]

Boundary definition is an ongoing process, not done once and for all but refined slowly over time with increased understanding of the system and its context. Usually, boundaries are set rather large initially and made progressively tighter. It is better to err on the side of slightly too large initial boundaries because it is easier to narrow the scope of an analysis than to enlarge it when its constraints and restrictions are experienced.

Fourth, do not feel compelled to do an exhaustive analysis to define a system's context. Do, however, consider organizational structure, culture and politics. Kling (1982) has examined the question of defining boundaries from the perspective of "analyzing the social dynamics of computing development and use in complex organizations." He has argued that the boundaries around a system should be drawn sufficiently large to include "the social and political *contexts* in which the computer-based system is embedded, the *infrastructures* for supporting system development and use, and the *history* of local computing developments." At the same time, one has to know where to stop. Kling jocularly points out that one could enlarge the context of a system "to include all human beings living and dead since the Ice Age in continental shelf territories and all artifacts ever built by these people which required the explicit cooperation of 7 or more people." He concludes, quite appropriately, that "there would be no explanatory gain" compared to a more reasonable definition of boundaries.

The initial definition of a system's context does not require extensive analysis. Setting the initial boundaries is something that a manager should be able to do with a little thought and no more training than that of reading the examples in this book. An internal or external consultant called in to help define the context should be able to ask the questions needed for an initial boundary definition in a two-hour interview. Participation of affected parties is probably not necessary to define the boundaries initially because the initial boundaries will inevitably be questioned and redrawn as system building proceeds. The purpose of boundary definition is to prevent premature constraints on solutions and design features. While this is essential, it requires the right perspective more than it requires time and energy.

CONSTRAIN THE DESIGN FEATURES

The second step in the design and implementation of systems from an organizational perspective is to constrain the design features of the solution. This means eliminating any designs that are infeasible for technical, cultural or, especially, political reasons. Political realities exert powerful constraints on the design features of systems. In fact, one reason that managers and system analysts rarely use the large definitions of context suggested earlier is their intuition about what is politically acceptable in their organization. In addition, a major danger of opening up the range of potential solutions is that expectations will be raised, later to be dashed. The trick is to explore the limits of the political constraints without creating false hopes.

Why Bother?

The primary reason for constraining the design features of the solution, whether system or organizational change, is to ensure that the solution matches the features of its context that should be preserved. This enables the solution to achieve its desired results without resistance or negative effects. Examples throughout the book have demonstrated the results of a failure to match the features of systems and structural changes to their context.

An equally important reason is to provide clear direction to the many people, both system professionals and users, who become involved in almost every system-building project. These people have differing interests and objectives that the process of system building gives them the opportunity to pursue. By clearly articulating the minimum requirement and the outer limits of the solution, a manager can avoid, at least in part, the pains of going back to the beginning. These pains are not merely those of delays and dollars down the drain but also of dashed commitment to a particular outcome on the part of those who assisted in its design.

An example from my personal experience illustrates the importance of properly constraining the design problem before turning it over to those who will fill in the details. Early in my career, I worked as a member of a team of internal consultants who were attempting sociotechnical redesign of a manufacturing facility. We believed that the nonmanagerial employees should participate in designing every aspect of the new facility in which they would be working, so we formed task forces of workers and managers to consider job design, rotation, training and progression from job to job and the structure (not the dollars) of the payment scheme.

The workers had been used to a payment plan that assigned a different base hourly rate to each partial-task job (clean-up person, labeler operator, bottle packer) and an individual incentive of up to 30 percent per day based upon the amount of production completed. Production for labeler operators was fairly easy to determine, but imagine how difficult it would be to determine objectively whether the clean-up person deserves a 30 percent bonus on a particular day. The workers in this plant believed the payment scheme was arbitrary and played games with managers both in the time studies on which the base rates were set and in the awarding of the daily bonus.

When the workers were given the opportunity to participate in the design of the payment scheme, their behavior surprised even the most hardened pessimists. Rather than trying to get every penny they could, they quickly decided upon a system that was simple, equitable and supportive of team work rather than competition. They proposed a no-bonus compensation plan in which their wages would be computed entirely from their hourly rate multiplied by the number of hours worked. The structure of the hourly rates was vastly simplified. There was to be a single base rate for trainees and then an increment for each of five core jobs learned. Thus, the highest paid worker would be a person who had made himself or herself most valuable to the team by cross training on every job. The base rate and increments were to be set in alignment with prevailing local wages, and then, to compensate for the loss of the bonus, the workers were to receive a flat 20 percent increase over their base wage. The average worker was making well over 25 percent bonus under the old system.

The team of internal consultants and the management of the redesigned plant were thrilled with the proposal until they tried to clear it with authorities from corporate industrial relations, some of whom had participated with the task force in generating the proposal. This group believed, quite reasonably, that other plants that were not redesigned and not working in a team fashion might demand similar benefits without being willing to work in a more responsible fashion. The proposal, therefore, was not approved.

The task force members were very upset and frustrated by this turn of events. They had been led to believe, by the formation of the group and the participation of industrial relations specialists, that they would have an influence on the outcome; and they had believed it. When their proposal was turned down, they felt deceived, and their enthusiasm for future compensation design work was considerably dampened. The entire fiasco could have been prevented if there had been a clear initial statement of what kinds of solutions would or would not be acceptable to corporate industrial relations.

Why Can't the System Analyst Do It?

There are two answers to this question. First, the commitment that a potential solution is in fact feasible can only be given by someone who has managerial control over all units in the organization affected by it. The manager of Grant Electronics IBD did have managerial control over solutions that involved all sections of her department but not over those involving the interaction of the IBD with payroll or of the human resources development division with the other parts of the organization. To pursue these higher-level solutions, the manager of IBD would have had to enlist the active support of the manager of the human resources development division in the first instance and probably the president of Grant Electronics in the second. Failing this active support, the manager of IBD might have succeeded in implementing higher-level skills through the exercise of superb political skills (a strategy potentially hazardous to one's career).

The second answer to the question is that system analysts simply are not trained to perform these tasks. Their methods do not require them to consider solutions other than systems or to challenge the problem definitions given them by managers on any grounds other than technical feasibility. (Nevertheless, they make good use of the few tools at their disposal, as chapters 6 and 7 showed.)

The exception that proves this rule is the ISAC (Information Systems Work and Analysis of Changes) method of system building developed by a group of researchers at the Royal University of Stockholm in Sweden. Developers of the method argue that change analysis must precede efforts to build information systems:

> The work that precedes information systems development is called change analysis. If the change analysis leads to the conclusion that the needs and

problems of the users are of a different kind than the information systems kind, other development is performed instead. . . .

The purpose of change analysis is to study what types of changes (improvements) that you should strive for in order to do something about the problems and needs you experience in the activities of your organization. It is important to try and find the reasons behind the problems and not only the symptoms. It is further important to work with a suitable combination of development measures. An evaluation of social, human, and economical factors helps us in this. It is finally important that the combination of development measures together can be expected to give the desired results. . . .

We decide upon systems development as the suitable development measure only if the change analysis indicates that there are problems and needs in the information systems area. In other situations, we should choose other development measures, e.g.,

- development of direct business activities, e.g., production development, product development, or development of distribution systems
- organizational development
- development of personal relations (communications training). [Lundeberg et al., 1978: 27–28]

Lundeberg and his colleagues describe a carefully thought-through and tested system-building approach in which users and professionals first document the existing organization using the same graphical conventions that they later use to document the design features of the proposed computer-based system, if any. Among other activities, the design team analyzes "the interest groups involved and the problems and needs for changes that they experience" (1981: 106). Subsequently, the design team generates change alternatives for each problem area and evaluates these against human, social and economic criteria. These alternatives include structural changes in the organization, changes in reward system, and so forth, in addition to automated systems.

The major strengths of the method are its explicit recognition that, first, there are many good solutions to any class of problems and, second, computerized systems are only one of these. The major weaknesses of the method are two. First, it is time consuming because it requires the participation of more than a few individuals. Second, in practice, the change analysis is rarely performed, even by advocates of the method. Canning described a seminar at which the ISAC method was presented by system professionals who had adopted it. He noted that "at least the three users [in this case, system professionals] that addressed the seminar tended to skip over this activity analysis and instead go directly to information analysis" (1979: 8).

How to Do It?

Constraining the design features of the solution starts at the point at which the initial definition of context leaves off. Again, there is no neat algorithm, but the guidelines for the process follow directly from the interaction perspective.

First, identify the groups of actors within the initial definition of context. Identify their interests and objectives; What kinds of changes can they be expected to favor, to resist?

Second, examine the problems and opportunities as they appear within this context. What is causing the problems: incompatibility of goals, handoffs in the work flow that cross organizational boundaries, responsibility without authority and resources? Create lists of the incidents that typify problems and opportunities and generalize the underlying issues that give rise to them.

Third, generate a set of potential solutions that spans a full spectrum of changes within the defined context. Consider, for example:

· Better manual systems, as well as computerized systems;
· Changes to the processes by which the organization's human resource is managed (e.g., compensation systems and work design);
· Structural changes;
· Organizational development;
· Business strategy changes;
· No changes;
· Symbolic gestures.

Fourth, identify those potential solutions that fall inside and outside the domain of control of the people participating in this design process. Eliminate unacceptable solutions within the domain of control, after challenging the reasons for their unacceptability. Devise plans for enlisting the necessary support for solutions outside the sphere of control, or eliminate them from further consideration.

Fifth, describe those features of the existing context that the solution must match or reinforce. Eliminate potential solutions that cannot be made to fit with the existing context.

This procedure will serve to constrain the range of potential solutions and to specify the minimum design features for further design work. Like the initial definition of boundaries, this procedure need be neither detailed nor time consuming. Later stages, possibly including traditional system analysis, will refine the conclusions of this process, eliminating more alternatives and adding more design features. The entire exercise can probably be conducted in a two-day session with a few key decision makers and a trained internal or external consultant once the initial boundaries have been set.

SPECIFY THE DESIGN PROCESS AND ROLES

The third step in effective system design and development, from an organizational perspective, is to lay out a procedure to be followed for the remainder of solution development and to assign roles and responsibilities to the key participants, be they users, computer professionals or other specialists like organization development practitioners. The solution development procedure must address two interdependent issues: first, whether the solution will be fully specified before system building begins or whether evolutionary development or minimum critical specification will be used and, second, whether user input will be obtained in traditional ways or whether participatory or political approaches to solution development will be used. Once these decisions have been made, the appropriate roles for various users and professionals can be spelled out. However, the ability to carry out these roles may require changes in the infrastructure of system development, at least for the duration of the development effort.

Clearly, the roles and responsibilities allocated to users and professionals will depend upon decisions about the degree of design specification and the nature and extent of user participation. Less clearly, decisions about design specification and user involvement must fit earlier decisions that constrained the size of the solution set and that specified minimum critical design features. However, just as the design features of a system interact with its context to produce changes, the process of system building exerts an influence over the design features produced. Tiffany's lamp shades were not created on assembly lines, and systems and other solutions capable of achieving certain outcomes simply cannot be built using traditional development methods.

Why Bother?

The major reason for specifying the process and roles for solution development is to ensure that the system and other designed changes fit within the constrained set of solutions and possess the minimum critical design features. The second reason is that an inappropriate design process can generate problems quite separate from a bad solution—for example, falsely raised expectations, apathy and resistance to the implementors.

The literature on system development has consistently advocated the involvement of system users in the process of designing and developing systems. For example, Lucas has identified these benefits of user participation:

1. It is ego-enhancing and builds user self-esteem.
2. Participation can be challenging and intrinsically satisfying.

3. Participation usually results in more commitment to change.
4. As a part of the planning process, participation means the user becomes more knowledgeable about change and is better trained in the use of the system.
5. We can obtain a better solution to the problem because participants know more about the present system than computer department staff members.
6. Participation means the user has retained much of the control over operations. [1981: 109]

However, empirical evidence on the value of user participation is mixed. Ives and Olson (1981), for example, have reviewed studies on participation in the development of computerized information systems. They found no consistent findings across studies about the effects of participation on system quality, system usage, user attitudes and user satisfaction with the information received. Brownell (1981) reviewed the research on participation in budgeting and concluded that various moderating variables like cultural setting, organizational structure, reward schemes and personality variables may account for the mixed effects of participation on effectiveness.

Furthermore, some evidence suggests that participation may sometimes have harmful consequences. Argyris, for example, has explained that the intention to accept or consider seriously the recommendations of the people participating does not always accompany the offer of an opportunity to participate, what he calls "pseudo-participation":

> After the executive had told [the researchers] that he insisted on participation, he would then continue by describing the difficulty he had in getting the supervisors to speak freely. "We bring them in, we *tell* them that we want their frank opinion, but most of them just sit there and nod their heads. We know they're not coming out with exactly how they feel." [1957: 145, emphasis in original]

Discussing sociotechnical strategies of productivity and quality of work life improvement, Cummings (1974) points out that employee participation should not be used in every situation. In situations where conflict, suspicion and low trust exist between workers and managers, participative approaches will often by rejected by workers. If workers do become involved in the process, managers may feel threatened and thus may ignore or override the workers' suggestions. In short, the same participative approach that may build consensus in a setting of cooperation can increase conflict in a hostile situation.

Participation in system design gives users an opportunity to select design features that allow them to achieve their political objectives. This is why, for example, corporate accountants in GTC (chapter 4) did not allow divisional accountants to participate. The corporate accountants knew the features they wanted in FIS and knew that the divisional accountants would try to alter them. The corporate accountants gambled that they would be able to override any resistance once the system was built by pointing to the large investment in the design and the expense of changing it.

Some implementors, however, equally committed to an unpopular solution, will attempt a process of participation anyway, relying on their ability to manipulate the participants into recommending the preferred solutions of the implementors. People are often adept at recognizing attempts to manipulate them, and their usual response is to feel angrier and more resistant than if the implementor had presented them with a fait accompli and demanded that they live with it. This type of manipulation is similar to bribery and coercion, and:

> Without making a value judgment, the least one can say about [a situation in which system implementors attempt to bribe or coerce the users] is that:
> 1. The probability that the bribed or coerced party will cooperate voluntarily in the future is reduced.
> 2. The probability that he will comply voluntarily to the implications of the choice which is made is also reduced. [DeBrabander and Edstrom, 1977: 193]

In general, then, user participation in system design should not be used when:

· It is intended as a symbolic gesture,
· Low trust or high conflict exists among the participants,
· The outcome has already been defined and participation is intended as manipulation.

A decision to allow users to participate is, in effect, a decision to allow users to determine the design features of the solution within whatever constraints are specified, which is why it is so important to set the constraints and minimum features carefully. The way in which user participation is structured will influence the quality of the solutions designed as much as the models of systems analysis used by computer professionals. For example, participation can take place within the approaches of full design specification, evolutionary development or minimum critical specification, as discussed in chapter 5. In the first case, users must provide all of their input before a system or solution is built; in the second, users can provide input to the designers successively, as they gain experience working with it; in the third, the design is deliberately left incomplete so that users can experiment with various ways of organizing. Clearly, evolutionary development and minimum critical specification and more likely than full specification to enable participating users to design effective and innovative solutions to the problems in their work place.

However, some practical drawbacks are inherent in development strategies that rely heavily on user involvement. Limited user involvement, for example, obtaining user input through interviews or questionnaires, can be accomplished relatively quickly. Extensive participation, however, usually requires a lengthy process of training users to work together in a group and to make decisions that are customarily made by their bosses. This process may often be measured in terms of months and years rather than days or weeks.

Consider, for example, the sociotechnical approach to system design (Taylor, in press; Bostrom and Heinen, 1977; Bostrom, 1980) that, like the ETHICS method described in chapter 5, entails extensive user participation. Describing the sociotechnical design process, Taylor listed the strengths and weaknesses of the participative approach:

> In this writer's estimation, participation by representatives of all . . . work groups . . . added significantly to the relevance and quality of the analysis and the resulting recommendations. . . .

> [But] the researcher learns, in cases like this one, to surrender his pursuit of crisp schedules and short deadlines. [In press]

Taylor was referring here to the fact that the employee team took ten months to perform their analysis. They worked on the analysis part-time while continuing to work at their regular jobs. It then took an additional twelve months to implement the recommendations of the team in entirety.

> If employees are really to be involved in the investigation of their own organization, they must be given the time they need to learn (sometimes relearn), and to teach one another what they learn. [In press]

Clearly, however, this is not the approach to use when an organization is in crisis or when the boss is very impatient.

For situations in which participation should not be used for political reasons or cannot be used for practical reasons, there are still some alternatives to the traditional system development approach, discussed in chapter 5, that fall loosely under the heading of political approaches to system design. Few people advocate political approaches, perhaps because of their riskiness or because of the stigma of social undesirability. Also, most of the political advice given to would-be system implementors can better be described as tactics or rules of thumb than full-scale development approaches.

For example, Keen has addressed the problem of what to do about users who can be expected to resist a proposed system. He seconds political tactics recommended by Bardach (1977), including:

· "Creating substitute monopolies": if the users are powerful, the implementors must be, too. Have something they want. If they want information services and automation, use this fact to obtain your own objectives.
· "Coopting likely opposition early": identify those parties most able to resist successfully. Get them on your side, and let them help you fight other resistors. (This entails careful political analysis in advance.)
· "Providing clear incentives": offer a variety of desirable inducements in return for compliance with the specifics of this system. [Keen, 1980b]

Keen summarizes his political advice to system implementors in these points:

(1) Make sure you have a contract for change;
(2) Seek out resistance and treat it as a signal to be responded to;
(3) Rely on face-to-face contracts;
(4) Become an insider and work hard to build personal credibility;
(5) Co-opt users early. [Keen, 1980b]

Similarly, Pfeffer (1981) has identified a number of political strategies and tactics that can be effectively adapted to the implementation of systems:

· "The selective use of objective criteria": choose the set of numbers that make your project look good.
· "The outside expert": hire a consultant who will recommend your desired course of action.
· "Controlling the agenda": set up the structure of the decision making session in ways that avoid unfavorable debate on your project.
· "Coalitions": create internal alliances and strengthen your ties to external constituencies.
· "Cooptation" and "committees": identify potential adversaries and get them to play on your team. [Pfeffer, 1981, chapter 5]

In addition, Pfeffer described the role of political language and symbolism in achieving desired outcomes. Applying this to systems implementation, Markus and Pfeffer (in press) recommend that designers pay close attention to the images projected by systems and their physical trappings. It is interesting to speculate what would have happened had the corporate accountants at GTC called their system managerial accounting system or divisional accounting system instead of financial information system.

The majority of the political moves described so far may be more aptly described as tactics than as strategies. None of them approaches the sophistication of the participative methodologies employed by Taylor (1981), Mumford and Weir (1979) and others. However, a few writers have begun to articulate methods suitable for the development of systems in politicized environments.

Turner, for example, has described a bargaining model for achieving consensus in an organizational setting characterized by multiple, incompatible goals. The setting he described was Columbia University, with its loosely integrated colleges, laboratories and campuses. The system, a new student records system, cut across almost every administrative unit in the university. The difficulty of achieving agreement on system requirements in this case can be inferred from the facts that:

> The admissions function is decentralized; each school has its own admissions office, establishes its own procedures, and sets its own standards. Student financial aid is also decentralized. . . . The registration and grade recordkeeping activities, on the other hand, are centralized in one unit. [1981a: 143]

For this situation, a strategy based on multiple work groups was developed. Four groups were formed on the basis of function within the student life cycle: admissions, financial aid, registration, current records and inactive records. Forty-five academic administrators and system specialists became involved in the overlapping groups and in the parent committee to resolve key issues and coordinate the groups. Even obtaining this level of activity appears to have been a major feat in the university setting. "Many of the proposed task group members had never met each other, nor had any of them ever been through a system study before" (Turner, 1981a: 147). Furthermore,

> Since a direct appropriation from Central Administration for the project could not be expected until after completion of the requirements definition study, the work of the task groups had to be funded out of existing budgets. This requirement meant that the Registrar [who had taken a primary role in the project] had to negotiate with the Dean of each school in order to have a particular person assigned to one of the task groups. [1981a: 146]

The four work groups were given common deadlines and operating procedures. They were expected to describe and document the existing system, to evaluate the existing system and to identify requirements for the new system. Turner's evaluation is that, by and large, the work groups produced excellent reports within the established deadlines. In contrast, "the process consumed a lot of time and effort. If it had been funded as an appropriation, the project would have cost in excess of $100,000" (1981a: 148).

On the whole, this strategy differs from participatory approaches chiefly in its multiple work groups. Mitroff and Mason have advocated a quicker but more controversial and adversarial approach that is an outgrowth of an experience Mason had while consulting with a large equipment manufacturing and construction company:

> Two groups of equally powerful executives had diametrically opposing plans for the future development of their business. As a result of the basic conflict in plans, the groups were at an impasse. Neither group was able through rational agreement alone to persuade the other of the reasonableness, let alone the "truth", of its plan. Because the groups were roughly equal in power and status, neither was able to force its view through over the other. More frustrating still, data or facts by themselves were not able to verify or refute key claims in either argument. The reason is not that either case was impervious to empirical test, but because of the complexity, openness, and fluidity of real world arguments. As a result of this fluidity, each side was able to take almost any piece of evidence relevant to the issue and interpret it in its favor. Through intensive study of both groups, Mason was able to work backwards from the surface of the conflict (i.e., which of the two policies was best) to a deeper set of underlying issues. It turned out that these issues were the real source of the debate. Unbeknownst to themselves, both groups were making a host of assumptions about their internal managemental capabilities, the external market structure, and the general external social and political environment. While of critical importance, these assumptions were rarely articulated, let alone debated.
>
> Mason discovered that it was possible (1) to state the assumptions of both groups in a dialectic format, (2) that this kind of format was pertinent to the kinds of complex issues managers and policymakers faced, and (3) that it proved helpful in understanding the complexity of the issue and deciding upon its ultimate disposition. A dialectic format meant that for every assumption which underlay the policy of one group it was possible to construct a contrary or strongly antithetical assumption which underlay the policy of the other group. . . .
>
> Next, a joint meeting was arranged and the dialectic was presented to both groups simultaneously. The underlying assumptions of the policies of both sides were arranged directly opposite one another on a large display chart. The meaning of each assumption was then illustrated by applying it to a common bank of data items, that is, a set of data which both sides had previously agreed was relevant in some not fully understood sense to both of their policies. The purpose of this step was to illustrate the fundamental point that standing by itself the data that supposedly bore on the dispute had no intrinsic, let alone singular, meaning. The data only took on meaning by being coupled to a world view—in this discussion, a strongly related set of assumptions. [1981: 30–31]

In brief, the Mitroff and Mason strategy entails bringing together the parties with opposing interests and taking them through a structured process to identify the root causes of their conflict. Such a meeting, properly facilitated, may lead to the development of a solution acceptable to both parties in a way that is not possible using a participative approach.

In summary, the process of designing systems exerts an influence over the nature and quality of the system produced. By specifying the design process to be used and by allocating roles to users and system professionals consistent with the selected process, managers can insure that the systems built meet minimum performance criteria without exceeding constraints and without producing negative organizational impacts.

Why Can't the System Analyst Do It?

The manager, rather than a system professional, must accept the responsibility to specify the design process and allocate roles. There are two reasons for this. First, the infrastructure of systems in most organizations is designed to support only one role for system analysts: managers and controllers of system design and user participation in it. Expecting them to suggest alternative ways of operating is expecting them to act in opposition to their own interests as defined and reinforced by their organizational structure and reward systems.

Second, the ethos of system professionals and the tools and techniques they have developed do not support other modes of behavior. System analysts view their role as that of a professional, like doctors or lawyers. The doctor asks the questions and recommends the diagnostic tests. The lawyer asks the questions and decides whether or not the client should take the stand. The system professional manages the process of design and determines the type and extent of client involvement:

> The systems designer plays a role similar to that of an architect or artist who receives a commission, talks with the client, and returns to a studio to create the desired product. [Lucas, 1981: 105]

No matter how extensive the involvement of users, the role of the professional designer is to structure it, manage it and control it:

> The designers meet with a large group of potential users to discuss the general boundaries of a system. . . . A design team meets individually with users to develop an understanding of present information processing procedures. . . . The designers solidify their understanding of the present system. Users are now asked to design the output they would like to have. . . . Finally, the users work to develop a format for the output. . . . During this time designers and users meet in groups. [Lucas, 1981: 111–112]

The way in which system building is structured and managed in most organizations reinforces this conception of the system analyst's role. Recognizing the need for a more active role for users, some people have advocated that analysts be retrained in social science or in new methods of analysis. Retraining the analysts, however, is not likely to produce the desired results without corresponding changes in the structured roles and relationships between users and designers. For systems to be designed from an organizational perspective, users and their managers, rather than system professionals, must control the process of analyzing and designing system solutions.

Consider these comments by Langefors, one of the developers of the ISAC method, in critiquing an article on the sociotechnical systems approach:

> Bostrom and Heinen [1977] argue that the bad MIS designs are to be attributed to the way MIS systems designers view organizations. They conclude . . . that it is necessary to change system designers' perspectives [by training them in the socio-technical systems approach]. . . . My remedy . . . is not to try to make "systems designers" supermen who would be both computer experts and human science experts, but to assign to the so-called "systems designers" their proper role, and label them designers of that subsystem of "the System", which is the "data system". . . .
>
> My own view . . . [is] that one cannot solve the problem simply by changing the outlook of the technological experts. Instead, the various users . . . were entered as the main designers of the overall system; the technical people took the tasks of subsystem design teams. . . .
>
> If one agrees that the data system is a subsystem [rather than the entire system], then it becomes no longer necessary to conclude that the view of data systems designer is crucial to the system design. What is crucial is how they satisfy the specifications for their subsystem as developed [by the users]. . . .
>
> It does not follow that the technical specialist members of the design teams ought to become expert social scientists; they could hardly become such without becoming less competent as technical designers. . . . It is not any more possible to obtain a good design by letting social scientists dominate the design; social science is not good enough. Consequently, it has been my thesis that the users, all affected people, should be among the main system designers and must have an influential, probably dominating, position. [1978: 56–57]

Thus, Langefors believes that the critical element in the design of a good system is who controls the analysis and design. He argues strongly that users should take control and allocate to system professionals only those tasks that concern their technological expertise. The rationale behind Langefors's argument is that users have knowledge about their needs that no one else, social scientist or computer professional, can have. But there is another, even more compelling, reason to argue that users and managerial decision makers should take control of system building. Users and managers, on the one hand, and designers, on the other, are both parties to the political process of system design and have different interests or stakes in the outcome. If computer professionals control the process, they will try to advance their aims where these conflict with the interests of users and managers.

This argument was clearly stated by Wagner at a conference on the implementation of computer-based decision aids:

> These heretical views lead me to argue that managers, systems designers, and researchers must recognize a potential conflict of interest regarding implementation. The manager presumably wants to improve the economic performance of the enterprise. Given this responsibility, the manager is willing to put resources at risk if there is reasonable chance of a good payout. Hence, the manager may be amenable to initiating and fostering a systems design project, but certainly has and must exercise the prerogative of calling a halt, or ignoring the results, whenever the economic benefits [or organizational impacts] look questionable. Thus, implementation per se is not the manager's appropriate end goal in systems development.

> The designer, in contrast, is responsible for providing decision makers with systems approaches that yield economic benefits. But since the designer does not have the manager's decision-making responsibility, the designer's success can be measured only in terms of whether the system is accepted, and perhaps whether it has been developed efficiently. Curiously, the designer takes little solace when a company's profits increase because an executive makes a decision that is counter to what the new system indicates but actually proves to be correct.[1975]*

The implication of this analysis is that managers and users owe it to themselves to take control of the system-building process, a rather ironic reversal of the advice that implementation researchers have been offering for years to computer system developers; Obtain top management support.

How to Do It?

Like defining the context and constraining the solution, specifying a design process and allocating roles and responsibilities need not require a time-consuming effort (although executing the desired design process may). The

---

*Wagner goes on to say that academic behaviorial researchers have still a third, conflicting set of interests regarding implementation, but in keeping with my own interests, I have omitted his description of these from this excerpt.

guidelines for a system design and implementation strategy consist of a set of questions that build upon the context and the feasible solution set identified earlier.

First, determine whether participation is an option. Participation can succeed only when managers are willing to consider input seriously, when a relationship of relative trust or cooperation exists between managers and other participants and when the possibility of multiple solutions or multiple variations on a single solution exists. If only one solution is feasible—for example, organizational change is not possible but system building is—then decide if flexibility exists in the design features of the system. If only one design concept is feasible for technical or political reasons, forget user participation or restrict it narrowly to the areas in which their influence will be allowed. Concentrate on selling the solution or implement it by fiat. If the context is highly politicized, characterized by conflict or hostility, consider political approaches to solution design.

Second, weigh the trade-offs between the benefits of user involvement in a participatory or political design process against the constraints of time and urgency. How important is it to the success of the organization for people to become more responsible, more self-directed? How important is it that people understand and remember the reasons behind the changes made so they can identify when the solutions no longer fit the context? How important is it that people be able to initiate and design improvements in the work organization? In organizations with low turnover, high length of service, high labor costs and high standards of quality or service, the answer to these questions is probably "very important." In high turnover organizations with low performance demands, the answer is probably "not very."

Another consideration is urgency. The luxury of user participation may have to be foregone if the success of the organization depends upon immediate improvements. Participation requires time for learning and trial and error, which is not available in crisis situations. A related concern is the tenure of the managers of the unit: They are not likely to invest in a process whose payoff will come long after they've moved on to a new job.

With the answers to these questions in hand, the decisions about which variation of participation or political process to choose are relatively straightforward. Extensive participation approaches (e.g., ETHICS, sociotechnical system analysis) require lots of time but build self-renewing organizations. Evolutionary development and minimum critical specification achieve many of the benefits of extensive participation in shorter periods of time. Because they begin to produce results immediately, they may be useful in crisis situations where the full specification approach is as likely to fail as methods requiring extensive participation. The traditional, full specification approach is useful in high turnover situations and situations in which the users have incompatible goals because it does not require users to come to agreement with each other or with the system designer.

Third, identify those who are to participate, and designate their roles and responsibilities. The parties affected by the proposed solution are the logical candidates for participation or for the more limited role of input provider. These parties were already identified when the constraints on the system were set, but the list of participants should be reviewed and revised at this time. A project manager should be designated in the user organization, someone with sufficient credibility and power to influence the departments represented by participating users as well as the system infrastructure. Structure the relationship between the participating users and the system infrastructure in ways that ensure not only that the constraints on the solution features are respected but also that the desired implementation strategy is followed.

CONCLUSION

The organizational perspective on systems leads to the recommendation that managers who would build systems adopt an explicit and active role in problem definition, solution generation and implementation planning. Traditional methods of system analysis and design are based on the assumption that problem definition and solution generation have already been done, and they prescribe a single strategy of implementation for every context and system. Many system professionals do not have the perspective and training to know when these tasks have been poorly done, and most do not perceive their role as challengers of managers' requests for assistance. Consequently, only the managers who request systems can ensure that these critical steps are performed—and only by performing them themselves. Managers should first define the context, then constrain the solution's design features and, finally, plan an implementation strategy with a specified process and roles, all before turning over a request for a system to a vendor or a system professional.

The organizational angle on systems also suggests a number of recommendations for improving the chances of system success. Once a detailed system design has been produced using whatever design approach (or before a prototype system is declared operational), an organizational impact assessment should be performed. Described in chapter 3, such an assessment consists of comparing the system design features with the context and predicting likely areas of resistance or negative impact. This process will enable correcting the problems before they happen.

Second, managers can reduce some system-building problems by making the minimization of such problems a criterion when they select a strategy for system acquisition. System acquisition strategies, discussed in chapter 6, include in-house development, external purchase and do-it-yourself computing. None of these strategies is right for all circumstances, but for any context, one of them might create many more problems than another.

Third, managers can reduce system-building problems by helping to restructure the infrastructure. Some ways of organizing the infrastructure were described in chapter 7. While none of these is better than the others in every context, one may offer system-building improvements in a particular organization.

The organizational perspective on systems suggests many ways to improve the effectiveness of systems and the organizations that use them. None of these recommendations is foolproof and none is especially easy, but using them consistently will help to transform the bugs of systems and system building into features.

Appendix A:
Vendors

Hardware Manufacturers

This segment of the computer world originally arose to produce and sell the
hardware components that comprise a computer system: central-processing
units, memory, input-output and storage peripheral equipment and communications devices. Initially, the hardware sector consisted of two major,
quite distinct, types of vendors: computer manufacturers and component
manufacturers. "More aptly called computer assemblers, [computer manu-
facturers] offer general purpose computers for sale to the general public"
(Sharpe, 1967: 183). Component manufacturers made the logic modules, core
memories, disk drives, and so on, required for complete computing capability
and used these either to develop computer systems for sale or to sell as
standalone products. "The large number of such firms makes it possible for a
computer 'manufacturer' to manufacture very little, choosing to simply select
and assemble components manufactured by others" (Sharpe, 1967: 183).
Component manufacturers are also called "independent peripheral vendors,"
defined as producers who do not manufacture computer central-processing
units. Those manufacturers who produce some components but who also
purchase components from others and then configure the equipment into a

212

computer system are known as "original equipment manufacturers," or OEMs (Phister, 1979).

Over time, this simple distribution of labor in the computer world, in which computer manufacturers made central-processing units, component manufacturers made peripherals and OEMs configured systems, has changed dramatically. Independent peripheral manufacturers still exist, although many are trying to become full-fledged computer vendors by manufacturing central-processing units. OEMs still configure computer systems, but some also sell applications systems (software) along with the computer systems dedicated to them ("New Rivalry," 1980). Also, computer manufacturers have expanded into the manufacture of components, the configuration of systems, software development, processing services and professional services. In addition, an entirely new segment has arisen, the plug-compatible peripheral and mainframe (loosely, large computer) vendors. These manufacturers created first components, then central-processing units, that mimic at lower cost and/or higher performance the operation of the equipment produced by the dominant computer manufacturers such as IBM and Digital Equipment Corporation.

Minicomputer Vendors

Since computers were first introduced for commercial and industrial use, there have been dramatic decreases in their physical size and corresponding increases in their memory and performance. Electromechanical and electronic computing equipment was developed in the 1930s and 1940s to tackle military and scientific problems. At that time, few people (or companies, including IBM) saw commercial potential in the cumbersome, expensive machines designed and built in universities. Business needs seemed almost exactly opposite the needs that had inspired computer development: instead of lengthy, complex computations on small amounts of data, business data processing entailed large volumes of data input and output and small amounts of rather simple processing. These requirements were adequately handled by electromechanical punched card tabulating equipment, a market then dominated by IBM.

When two professors from the University of Pennsylvania were immediately able to sell their ENIAC computer to the Census Bureau, IBM and others awakened to the commercial potential of computers, and the mainframe business was off and running. The ENIAC had storage space for only about 200 decimal digits, but it weighed several tons and occupied several rooms. Today, computers with much more storage space and power are being fitted inside automobile dashboards and microwave ovens. After the vacuum tubes of the ENIAC were replaced by magnetic cores in about 1950, the decrease in physical size and space requirements and the increase in internal storage capacity, speed and reliability were so great that vendors heralded the second generation of computers. When integrated circuits began replacing cores in the 1960s, the third generation was announced. What had formerly consumed the better part of a large room was now reduced in size to several bulky cabinets. Since this time, miniaturization of electronic components (the evolution of large-scale integration, very large-scale integration and extremely large-scale integration) has continued to the point where computers with the memory and processing capacities of third generation computers can now fit atop a desk.

About the time that third generation computers were ushered in, however, computer people stopped reckoning the size of computers in terms of generations. In the mid-1960s, the minicomputer appeared. Not only were minicomputers smaller in physical size than the third generation machines, they were also smaller in internal storage capacity, in processing speed and in price ("Note on the Minicomputer Industry," 1979). Originally intended for scientific and industrial use, the minis were speedily adapted to business and commercial use as small business computers (SBCs).

SBCs are general purpose machines much like their larger counterparts, the commercial mainframe computers. They have a variety of uses. A large organization may, for example, use an SBC to perform administrative processing for a subunit, to perform a single dedicated application that does not need to be integrated with other applications or, in combination with other SBCs, to perform processing for the entire organization.

Microcomputer Vendors

In the early 1970s, the microcomputer appeared. The micro was to the mini as the mini was to the mainframe: smaller again in physical size, memory size, processing speed and capability and price. For a few thousand dollars, one could obtain a machine that fit on a desktop but that could perform almost all the computation an individual could desire—hence, the name "personal computer." While micros do not have the memory capacity and processing speed of minicomputers, they have far more power than the best programmable calculators. They have been enthusiastically greeted as personal support tools by scientists, engineers, electronic hobbyists and, more recently, managers.

Just as quickly as minis, the micros were converted to commercial use. Personal or desktop computers are used by individuals for home use, by small businesses for administrative applications and by medium- and large-sized organizations for a variety of standalone applications. However, organizations have not widely integrated these devices into their existing data-processing applications or computing systems.

The vendors that initially developed and produced mini- and micro-computers were not the same vendors that had pioneered the development of mainframes. But many of the surviving mainframe manufacturers have expanded their product lines to include smaller computers. Industry experts believe that mainframe computation is a mature industry. Almost every organization that needs large computers already has them, and only a replacement market remains. Microcomputers are, however, an embryonic business, and it will probably be years before the market for them is saturated.

Plug-Compatible Vendors

Initially, computer hardware manufacturers focused their energies on the central-processing unit, the heart of a computing system. Other companies found a booming market for the add-ons that make central-processing units into systems, and for a while, independent peripheral shipments looked strong. When the computer companies realized the business potential of this segment, they began to manufacture their own peripherals. The shipments of the independents fell off, reducing their markets to users and independent system integrators (OEMs and systems houses) (Phister, 1979).

The advent of third generation computer mainframes, in about 1965, promised to stabilize many aspects of the technologically turbulent industry. In particular, IBM introduced standard electrical interfaces among its various pieces of equipment, making it possible for a customer to convert from one IBM tape drive to another almost as simply as unplugging one and plugging in the other. This innovation offered the independent peripheral manufacturers an opportunity to recoup their losses. They introduced families of plug-compatible equipment, which they marketed directly to IBM's customers as replacements for the peripherals obtainable from that vendor. The plug-compatible vendors offered lower prices and often also improved the performance of their equipment.

After 1970, a new plug-compatible market opened up: Vendors began to produce central-processing units that could serve as higher-performance-but-lower-cost replacements for IBM mainframe computers—in a phrase, plug-compatible mainframes (PCMs). These computers were designed to run under IBM's operating system software so that computer users could install them without massive and expensive conversions of their applications software. Such conversions would have more than eliminated any saving from better performance.

Much to the surprise of IBM, which had apparently ignored the PCM threat initially, PCM manufacturers were able to capture a significant share of business at the high profit end of its computer line. The PCM vendors prospered in spite of new product announcements by IBM and the uncertain legal status of their free ride on IBM software.

Leasing Companies

Another segment of the computer world grew up around the mainframe manufacturers, the independent computer-leasing business. The nonbank computer-leasing business exists principally for mainframe computers, necessitated by their originally very high prices. Hardware costs have gone down across the board, and the more affordable mini-based systems with their shorter product life cycles have not encouraged the expansion of the computer-leasing business into this area. Many people believe that this industry is waning because of the maturity of the mainframe market.

The Telecommunications Sector

Telecommunications refers to the movement of voice, images (still and video), data and text over electrical and electronic communications media such as telephone wires, coaxial cables, microwaves, optical fibers and satellite transmissions. Any of these media is capable of carrying all types of communications traffic (voice, image, data, text), provided that certain equipment, software and services are used. Whenever data must move more than a few hundred feet (the maximum distance over which peripheral devices can be directly attached to computers via coupling cables without data distortion), equipment, software and services for data communication must be used.

The telecommunications marketplace, encompassing as it does telephone, radio, TV and other media, goes far beyond mere data communications and, in fact, is much larger than the computer world. Equipment is the least significant part of the telecommunications arena. Vastly more critical to the development of information and communication systems are the networks that distribute the data (voice, images, text).

Who owns and operates telecommunications networks used in a computing system can strongly influence the hassles facing organizations. Publicly owned networks are owned, operated and maintained by vendors for use by any individual or organization. Examples are the networks operated by the telecommunications common carriers like AT&T and the specialized common carriers like MCI. Networks may also be owned in a shared fashion by several similar organizations. Banks, insurance companies and airlines, for example, have developed shared telecommunications networks. Large organizations with major telecommunications needs may find it advantageous to configure and operate their own networks from equipment and services purchased or leased from a variety of vendors. Eventually, many vendors will offer turnkey (fully configured, dedicated, ready-to-run) telecommunications networks. Today, however, most organizations find that they must design and configure networks in-house.

Computer System Marketers

One of the most interesting features of desktop computers is the methods companies have developed to market and distribute them. Because of their two distinguishing features—"they are cheap enough for individuals to buy and simple enough for many laymen to operate" (Uttal, 1981: 84)—desktop computer systems are mass marketed and distributed in volume. Like cameras and calculators, they are sold by licensed dealers who may also sell other products, by manufacturer-owned retail outlets and even by mail. In contrast, the mainframe vendors employ armies of salespersons and field engineers to attract potential customers and to give them the on-site personal attention and service associated with a multi-million-dollar product. This type of marketing is simply too expensive for equipment with small price tags.

Even the larger SBCs fall prey to this trend. Authorized dealers and distributors and retail business computer centers have sprung up to assist in the marketing of these systems. Overall, the trend toward the mass merchandising of computer systems puts a great deal of pressure on more traditional distribution channels like the manufacturers' direct sales forces and OEMs and systems houses.

VENDORS OF APPLICATIONS

Processing Services Firms

Processing services firms supply raw computer power, access to specialized data banks or processing of customers' data through specialized applications programs. Some processing services firms began as organizations in other lines of business, like banking, that wanted to sell excess computing capacity. However, most computer service bureaus were independent companies from the start.

By 1966, there may have been 800 service bureaus. These companies purchased or leased computer equipment and hired staffs of programmers, analysts and computer operators to provide companies with an alternative to setting up computing departments. Customers typically sent raw data (e.g., the weekly payroll) to the service bureau via mail or messenger. There, the data were keypunched or keyentered, assigned a priority (e.g., first come, first served), and then "run to completion on the computer with a proprietary software package, and returned to the customer and his employees in the form of summary reports and payroll checks" (Bower, 1973: 540).

Time-sharing technology, associated with third generation computer hardware, became available around 1966 or 1967. The time-sharing segment of the computer services industry arose soon after. Customer access to the computers of time-sharing service providers is accomplished through remote terminals and communication lines rather than through mail or messenger. While conventional service bureaus may still comprise the largest segment of the services industry, the on-line service bureaus or computer utilities have been the most rapidly growing segment of the industry for years and may soon overtake conventional service bureaus (Phister, 1979).

Today, customers of processing services firms can gain access to their services in any of three ways: by mail or messenger (which implies batch-processing computer technology) and by two on-line access modes—remote batch and interactive. The remote batch access method typically involves a customer company keypunching data at its own site and submitting the data to the bureau via a card reader and telecommunications lines. The results of the processing are returned to a line printer at the customer's site after a delay for processing. This differs little from the situation in which the input and output devices are attached to the customer's own batch-processing computer. In the interactive method of access, the customer enters data through a keyboard, either a printing terminal or a CRT display.

In addition, customers can make use of the services of these vendors in three ways: by purchasing raw computer power for in-house programming, by performing regular calculations with the vendor's software and by accessing common data files with the vendor's software (Phister, 1979). In the first mode of use, the customer writes all the software programs required to accomplish desired tasks; only the computer power to run these programs is acquired from the vendor. Some time-sharing vendors offer several high-level languages to enable customers to write their own programs.

In the second case, far more common, customers avoid the problems of developing software by using vendor-supplied programs. Historically, small companies, too small to afford their own computing facilities and staff but large enough to benefit from computerization, formed the customer base for this huge component of the market. The predominant applications were payroll, billing, accounts payable and accounts receivable. In this mode of use, vendor-owned software updates the customer's private data files. While precautions are taken to ensure that data can only be accessed by the proper customer, concerns about privacy and confidentiality arise. Consequently, as affordable mini- and microhardware becomes widely available, this customer market may contract.

In the third mode of use, customers use computer services interactively to access vendor-maintained, common data files. The first on-line services, airline reservation systems and stock quotation services, were of this type. Other examples include U.S. Census files, consumer credit files, entertainment ticket files and various specialized data bases, such as Value Line, Compustat, Dun and Bradstreet, U.S. Prices Data Bank and Ohio College Library Center's bibliographic data base.

Processing services companies have begun competing in nontraditional arenas. Not only are they selling services based on applications software, but also they are increasingly acquiring licenses to market these software packages independently of their own processing. In so doing, they compete with software houses. Processing services also have begun selling hardware and turnkey systems.

Software Houses

Software firms sell software packages. This sector of the computer world addresses those organizations that are willing and able to acquire, configure and operate their own computing equipment but unwilling or unable to develop their own software. The software sector has evolved into a two-tier market for a variety of historical and economic reasons. At the high end of the market is software expensively custom designed or purchased and modified before use. Ongoing maintenance and customer support (e.g., training) of this software forms a major item of expense. At the low end of the market are the inexpensive software packages one buys in retail stores. This software is often used without modification. Customer support is obtained by reading manuals and joining user groups.

The high end of the software market developed because the manufacturers of early computers sold them without the software necessary to make them something more than "a useless collection of electronic and electromechanical components" (Dolotta et al., 1976: 87). What is thought of today as the craft of programming was more an art than a craft in those days. Programmers had to teach themselves what sequences of instructions would accomplish what tasks without the benefit of schools, textbooks, higher-level languages or programming aids common today. "The IBM 701 . . . which was

introduced in the early 1950's, had only 32 instructions and the manual for it
was 103 pages long" (Dolotta et al., 1976: 89–90).

The instructions for each model of computer differed in form and in
function; what worked one way on this machine worked another way on that
one. The concepts of modular software (layers of self-contained boxes of code
that plugged neatly into one another) had not been conceived. Rules of good
programming practice were not formulated or written down until the advent
of structured programming in the early 1970s (Welke, 1980).

Another factor contributing to the inefficiency of programming was the
fact that each program had to contain lines of code for every subtask in it.
Today, repeatedly used functions are separated out into systems programs and
utility software. In the early days, however, every program written was an
application program (Dolotta et al., 1976), a very labor-intensive process.
Consequently, all software was custom made. The software package—that is,
multiuse or general purpose software requiring little or no tailoring—was
infeasible in the technological environment of the early 1960s. Software
simply was not transportable from one machine environment to the next.
Some computer hardware vendors, like IBM, offered programming assistance
to their customers (a practice IBM discontinued in 1969), but most believed
this activity was peripheral to their major functions. The necessity to undertake the uncertain activity of software development undoubtedly deterred
some potential computer users.

Third generation computer technology made portable software feasible. Several entrepreneurs began marketing software products in the mid-
to-late 1960s. The majority of the early products were systems programs
rather than application programs. Many early software vendors ran into
economic problems caused by the software pricing policies of the computer
manufacturers. When these policies were changed, the software products
market boomed. By the late 1960s, computer manufacturers had recognized
that unavailable applications software would constrain hardware sales because
not all organizations were willing to undertake in-house development.

Computer manufacturers tried several ways to increase the supply of
software. One was to develop and maintain directories of programs, developed for their equipment by other vendors or users. These vendor-
maintained directories tended to contain out-of-date or missing information
(Beeler, 1980), however, so the hardware manufacturers also tried, second, to
encourage sharing of useful programs directly among computer users. The
frequent meetings of user groups were intended to accomplish this sharing.
Third, computer manufacturers tried to encourage third parties to develop
software for their machines. They let contracts to programming firms for the
development of assemblers, compilers and various utility programs (Phister,
1979). Eventually, this practice fell into disuse, but other symbiotic relation-

ships between hardware manufacturers and independent software vendors exist, like joint marketing agreements. Fourth, computer manufacturers began developing their own systems software for sale.

Packaged software today is marketed by brokers, management consultants, accountants and computer store dealers, in addition to computer vendors and software firms. Software brokers act primarily as go-betweens (Sharpe, 1967). Some are staffed to provide installation, support and maintenance of the packages they sell; others defer these functions to the software developers. Management consulting firms sometimes offer to sell proprietary software, which requires their consulting skills to customize and use it:

> [A]ll but two of the 15 largest Certified Public Accounting firms are in [the software] business, sometimes selling the product directly and other times providing it as part of their auditing services. [Welke, 1980]

Likewise, computer stores initially appealed to the computer hobbyist, but today they are aimed more heavily at the business market and home use. No estimates are available for the anticipated volume of software sales through these outlets. However, even the hardware vendors used to more conventional marketing arrangements have started to retail their smaller systems and associated software.

Systems Houses

System houses produce packages of integrated computer systems and applications software. Historically, OEMs were the only source of integrated minicomputer systems. A potential buyer's alternative to buying from an OEM consisted of purchasing a mini-central-processing unit from the manufacturer, attempting to buy or assemble needed peripherals, building the necessary software and hoping the resulting system worked. OEMs stepped into a breach in the marketplace by manufacturing peripherals and selling configured systems.

Over time, as the hardware manufacturers began manufacturing peripherals and marketing systems, OEMs were pushed into residual roles that avoided direct competition with the major vendors. Some OEMs began building peripherals for specialized industrial applications where the market was too small to interest the hardware vendors. Others, particularly in the 1970s, began to perceive their greatest advantages over the manufacturers to lie in their ability to provide software tailored to a specialized industry group.

For years now, many of the 4,500 system houses in this country have been thriving on the development of turnkey applications ("New Rivalry,"

1980). However, the two dominant economic trends in the computer industry—the rising cost of software and the falling price of hardware—have had a paradoxical effect on this group of vendors. The demand for turnkey systems has risen enormously, even among larger users desiring the low maintenance, self-contained systems. The increased demand attracted competitors from among computer manufacturers, service companies and software houses. The system houses may discover that they can no longer afford to provide their own distinctive competence and customized software.

The Professional Services Sector

Professional services include custom programming, facilities management and consulting. Contract programming firms sprang up to assist computer buyers who had a hard time attracting computer professionals. Welke (1980) pinpoints the origin of these firms to 1959 and credits their formation largely to the needs of the U.S. government. For the next ten years, contract programming firms were the only real alternative to in-house software development. During the 1970s, the packaged software industry has overshadowed contract programming, but contract programming firms still occupy an important place in the computer world (Pomerantz, 1981).

Two kinds of contract programming firms can be identified today. One is the project management firm, which contracts with a computer-using organization to supply and manage the programming talent required to perform a defined piece of work—for example, developing an application program or performing an operating system conversion. Usually, the methodology for managing the software development process is a large part of the marketing appeal (and overhead) of these firms. The second type of firm contracts only to supply programming talent for variable time periods at hourly rates, much like temporary secretarial services. In this way, firms with project management capabilities can augment their permanent data-processing staff or fill temporary vacancies. The temporary firms are disdainfully referred to as "body shops" by the project management firms; the temporary agencies claim that they offer exactly what their clients want, laborpower without overhead. Perhaps the most significant difference between the types of firms is the fact that programmers are employees of the project management firms and therefore have some career development potential. Programmers working as free agents for the temporary agencies earn higher pay at the cost of low job security and advancement potential (Pomerantz, 1981).

Facilities management firms first appeared in the late 1960s. These vendors designed their services to bail out the organizations that had invested heavily in computer equipment before they found themselves unable to recruit, retain and manage computer professionals. These services have been especially popular in hospitals and educational institutions, but this firm type has not expanded as rapidly as other parts of the computer world.

The professional services sector also includes numerous independent consultants, research firms, information clearinghouses, publishing firms and educational institutions.

This brief review has only hinted at the breadth, variety and flux of the computing world. Its continually changing outlines, shaped by new products, services and firms, make a fascinating study in organizational behavior.

Appendix B: Evolution of the Computing Infrastructure

It seems remarkable that a field not much more than thirty years old should have its own historiography, but *The Annals of the History of Computing* began publication in 1980. Computer hardware is categorized in generations, and computer managers have been offered stage theories of growth (see Gibson and Nolan, 1974) to help them identify and cope with the critical issues facing them. However, the issues confronting computer managers and users have changed over time and are likely to continue to do so, suggesting the value of a review of business computing in the past thirty years and a projection into the next few years.

Vitalari (1978) has organized a review of historical computing trends into the three decades of the 1950s, the 1960s and the 1970s. The following discussion examines three slightly different decades—1954–1964, 1965–1974 and 1975–1984—but, otherwise, follows Vitalari quite closely.

1954–1964

This was the era of second generation computer technology. Vacuum tubes had been replaced by magnetic core memories so that the machines were reliable enough to be used for business purposes, but software development technology was primitive at best. Programs were not portable from machine to machine; there were no packages, no common applications and very little systems software. New technological developments required converting applications as well as equipment. Data were stored on cards, and little thought was given to the costs or benefits of data redundancy. Economics centered on the hardware: The key concerns were selecting the right equipment and using it most efficiently. The labor of programmers was freely expended to economize on machine usage.

The dominant philosophy of those who worked with computers in this period was automation: The replacement of clerical functions. The first applications were in the accounting area, where office work had already been studied with the tools of industrial and methods engineering and automated with electric accounting machines.

The organizational issues at this stage of computing were minor (as perhaps they always are for new computer users). Relatively little job differentiation had occurred, not much beyond analyst, programmer, operator and keypuncher. The major concern in the literature of the era was whether or not locating the computer in the accounting department would hinder its application to manufacturing and sales. The promotion of the head of computing to corporate vice-president was optimistically predicted but had not occurred. The benefits of centralizing computer processing within an organization were questioned as early as 1955, but the issue was moot in the absence of technology powerful enough to handle all of an organization's processing. Computers remained where they entered the organization (in accounting), and so did those who managed them.

1965–1974

All of this changed quite rapidly with the advent of the IBM 360. Not only did this machine represent a technological advance, but also it changed the way people thought about computing. It was a symbol as well as a machine.* The IBM 360 was rumored to stand for 360 degrees, full circle, the machine big enough to accomplish both scientific and business computation and permanent enough to mean the end, forever, of applications software conversions. Whether or not it delivered on these promises, the 360 sold very, very well.

The philosophy of applying such a grand tool was equally grand. The large capacity of the machine allowed previously separate applications to be

---

*Cynics and critics have argued that the 360 was more symbol than substance, merely popularizing technological innovations made earlier by other vendors. Its systems software lagged delivery of the machine by two years and never quite lived up to expectations. Nevertheless, it was and still is a spectacularly successful symbol.

integrated (at no little conversion cost), and data base software was created to support the integration of applications with the proper amount of data redundancy. It was at about this time that the term *management information systems* was coined to convey that computers could be applied to the real problems of management, not merely to clerical automation. The shibboleths *total systems* and *integrated systems* hinted that soon the computer would provide all the information managers need to do their jobs.

The power of the new machines allowed most users to centralize hardware processing at a single site. The price tags of the machines were so low that opportunity had the force of command. Special air-conditioned rooms with glass fronts were constructed to house the idols; their attendants now achieved visibility and status. In keeping with this awesome display of power, the high priests of computing were finally elevated to the vice presidency. This had payoffs for the vendors, especially Big Blue (IBM, because the computer cabinets of this vendor have always been blue in color), since a single powerful customer made for an easier selling job. Grateful MIS managers make the vendors' sales job even easier by establishing single-vendor shop policies. Throughout this period, the utility of computing to management was regularly proclaimed, but all attention was on the machine.

Hype comes before the fall. The period had not ended when managers became alarmed at the vast amounts spent and the paltry returns received. The phase of MIS control had begun. Steering committees were formed to ensure the accountability of centralized data-processing shops. Spending was subjected to cost-benefit analysis and, occasionally, postaudit evaluation. Standards were developed about every conceivable aspect of operation and program development. In fact, it is perhaps the highest achievement of this period that the development of applications software was transformed from an uncertain, error-prone process into a routine activity with predictably good, if much delayed and far too expensive, results.

1975–1984

SBCs, distributed processing and desktop computers have arrived on the scene, with remarkably little of the fanfare that used to accompany predictions of applications that never quite were. Computers are casually advertised on television by stars who candidly admit that they don't know what to do with them. Word processing is commonplace, and teleconferencing is starting to be used by business people.

This new type of computing power has a very different character from the computing of the middle decade: It is personal, portable, distributed and dispersed. Demand for small-scale computing has exploded at a time when the data-processing shops of the largest organizations have finally converted most basic business operations into systems that run on large-scale computing systems. The marketing philosophies of vendors are shifting to accommodate a view of computing as a personal support tool for management decision making. The term *decision support* is competing with MIS for status and is winning. At the same time, computer companies are making sales calls on managers directly rather than on data-processing experts, and they are opening retail stores and advertising computers on television.

This hardly portends the dismantling of the massive systems that keep the business running while managers make decisions, but people are nevertheless thinking about these operational systems in new and different ways. Instead of total, integrated systems, operational systems comprise a collection of deintegrated, selectively decentralized systems, some of which are monumental daily number crunchers, others of which are one-time-only analytic models.

The proliferation of tiny tools has coincided with the appearance of data base management software, inquiry languages and end-user programming. This coincidence has highlighted major differences in orientation between large- and small-scale computing. Large-scale computing maintains the organization's permanent records; small-scale computing processes different kinds of data including text, voice, images and some numbers derived from the core record-keeping systems.

These differences have effectively refocused attention from the machines onto the stuff they store, process and transport. "Managing the data resource" has become the new MIS slogan.* Data-processing shops have been forced to rethink their role from custodians of the machine to that of providers of the basic computing resource and custodian of corporate data. Thinking the unthinkable, some MIS visionaries predict the day when what remains of central data processing will no longer develop applications. Users will write their own programs as easily as they operate their automobiles, once a data-processing utility is available for them to use. When this happens, many people believe computing and its infrastructure will finally address the real needs of system users.

---

*One of the latest slogans, now the acronym IRM, is information resource management (see Matlin, 1980).

References

Ackoff, R. L. "Management Misinformation Systems." *Management Science* 14(4) (1967): 147–156.

"A New Era for Management." *Business Week*, April 25, 1983: 50–86.

Apcar, Leonard. "Angry Computer Users Sue Burroughs for Snags." *Wall Street Journal*, October 17, 1980a.

———. "Burroughs Strikes Back at Militant Customers." *Wall Street Journal*, November 25, 1980b.

Argyris, Chris. *Personality and Organization.* New York: Harper & Row, 1957.

———. "Resistance to Rational Management Systems." *Innovation* 10 (1970):28–35.

———. "Management Information Systems: The Challenge to Rationality and Emotionality." *Management Science* 17(6) (1971):B-275-B-292.

Bardach, Eugene. *The Implementation Game: What Happens after a Bill Becomes Law.* Cambridge, Mass.: MIT Press, 1977.

Bariff, Martin L., and Jay R. Galbraith. "Intraorganizational Power Considerations for Designing Information Systems." *Accounting, Organizations and Society* 3(1) (1978):15–27.

Becker, Franklin D. *Workspace: Creating Environments in Organizations.* New York: Praeger Publishers, 1981.

Becker, Franklin D., and Charles C. McClintock. "Mixed Blessings: The Office at Home." *Proceedings of the 1981 Office Automation Conference*, Houston, Texas.

Beeler, Jeffrey. "Applications Software Tops User Group Issues." *Computerworld*, June 16, 1980.

Benbasat, I., and R.N. Taylor. "The Impact of Cognitive Styles on Information Systems Design." *MIS Quarterly* 2(2) (1978):43–54.

Berger, Paul, and Franz Edelman. "IRIS: A Transactions-Based DSS for Human Resources Management." In Eric D. Carlson, ed., *Proceedings of a Conference on Decision Support Systems Data Base* 8(3) (1977):22–29.

Berrisford, Thomas, and James Wetherbe. "Heuristic Development: A Redesign of Systems Design." *MIS Quarterly* 3 (1979):11–19.

Bjork, Lars E. "An Experiment in Work Satisfaction." In Louis E. Davis and James C. Taylor, eds., *The Design of Jobs*, 2nd ed. Santa Monica, Calif.: Goodyear Publishing Co., 1979, pp. 219–227.

Bjorn-Andersen, Niels, and Poul Pedersen. "Computer Facilitated Changes in the Management Power Structure." *Accounting, Organizations and Society* 5(2) (1980):203–216.

Blundell, William E. "The Software Kid Inveighs Against the Witch Doctors." *Wall Street Journal*, July 6, 1981.

Blustein, Paul. "How the JWT Agency Miscounted $24 Million of TV Commercials." *Wall Street Journal*, March 30, 1982.

Bostrom, Robert P. "A Socio-Technical Perspective on MIS Implementation." Paper presented at the ORSA/TIMS National Conference, Colorado Springs, November 1980.

Bostrom, Robert P., and J. Stephen Heinen. "MIS Problems and Failures: A Socio-Technical Perspective." *MIS Quarterly*, 1 (1977): 17–32.

Bower, Richard S. "Market Changes in the Computer Services Industry." *Bell Journal of Economics* 4(2) (1973):539–590.

Braverman, Harry. *Labor and Monopoly Capital: The Degradation of Work in the Twentieth Century*. New York: Monthly Review Press, 1974.

Brownell, Peter. "Participation in Budgeting, Locus of Control and Organizational Effectiveness." *The Accounting Review* 56(4) (1981):844–860.

Bulkeley, William. "Variety of Desk-Top Computers Is Aimed at Fast-Growing Small Business Segment." *Wall Street Journal*, June 19, 1981.

Burlingame, John F. "Information Technology and Decentralization." *Harvard Business Review* November/December 1961:121–26.

"Burroughs Found Guilty of Fraud, Breach of Contract." *Wall Street Journal*, July 7, 1981.

Bylinsky, Gene. "A New Industrial Revolution Is on the Way." *Fortune*, October 5, 1981, pp. 106–114.

Canning, Richard. "The Analysis of User Needs." *EDP Analyzer* 17(1) (1979).

Chamot, Dennis, and Michael D. Dymmel. "Cooperation or Conflict: European Experiences with Technological Changes at the Workplace." Washington, D.C.: Department for Professional Employees, AFL-CIO, 1981.

Chandler, Alfred D., Jr. *The Visible Hand: The Managerial Revolution in American Business.* Cambridge: The Belknap Press of Harvard University Press, 1977.

Chase, Marilyn. "IBM Agrees to Provide Support Services for Software to Users of Other Computers." *Wall Street Journal,* June 12, 1981.

Cherns, Albert. "The Principles of Sociotechnical Design." *Human Relations* 29(8) (1976):783–792.

Conrath, David W. "Organizational Communication Behavior: Description and Prediction." In M. Elton and W.A. Lucas, eds., *Proceedings of the NATO Symposium on the Evaluation and Planning of Interpersonal Telecommunications Systems.* London: Plenum Press, 1978, pp. 425–442.

Conrath, David W., and Gabriel du Roure. "Organizational Implications of Comprehensive Communication-Information (I-CS) Systems, Some Conjectures." Aix-en-Provence: Institut d'Administration des Entreprises Centre d'Etude et de Recherche Sur Les Organisations et la Gestion, 1978.

Cougar, J. Daniel; Robert A. Zawaki; and Edward B. Oppermann. "Motivation Levels of MIS Managers versus Those of Their Employees." *MIS Quarterly* 3 (1979):47–56.

Crozier, Michel. *The Bureaucratic Phenomenon.* Chicago: University of Chicago Press, 1964.

Cummings, Thomas G. "Intervention Strategies for Improving Productivity and the Quality of Work Life." Paper delivered at the Eighty-Second Annual Convention of the American Psychological Association, New Orleans, Louisiana, August 1974.

Cummings, Thomas G., and Suresh Srivastva. *Management of Work: A Socio-Technical Systems Approach.* Kent, Ohio: Kent State University Press, 1977.

Dagwell, Ron, and Ron Weber. "System Designers' User Models: A Comparative Study and Methodological Critique." *Communications of the ACM,* in press.

Davis, Gordon B. "Strategies for Information Requirements Determination." *IBM Systems Journal* 21(1) (1982):4–30.

DeBrabander, Bert, and Anders, Edstrom. "Successful Information System Development Projects." *Management Science* 24(2) (1977):191–199.

Dolotta, T.A. et al. *Data Processing in 1980–1985: A Study of Potential Limitations to Progress.* New York: Wiley, 1976.

Emmett, Ralph. "VNET or Gripenet?" *Datamation*, November 1981, pp. 48–58.

Flamholtz, Eric, and Ann Tsui. "Toward an Integrative Theory of Organizational Control." Working paper 14. Los Angeles: UCLA Pacific Basin Economic Study Center, 1980.

French, John R.P., Jr., and Bertram Raven. "The Bases of Social Power." In D. Cartwright, ed., *Studies in Power*. Ann Arbor: University of Michigan, Institute for Social Research, 1959, pp. 150–167.

French, Nancy. "Computers Hit for Killing Jobs." *Computerworld*, October 13, 1980.

Galbraith, Jay R. *Organization Design*. Reading, Mass.: Addison-Wesley, 1977.

Gane, Chris, and Trish Sarson. *Structured Systems Analysis: Tools and Techniques*. New York: Improved Systems Technologies, Inc., 1977.

Garson, Barbara. "Overload in the Data Cluster: Vignettes of Life in the Electronic Office." *San Francisco Examiner This World*, June 28, 1981.

Gasser, Leslie. "The Social Dynamics of Routine Computer Use in Complex Organizations." Ph.D. dissertation, University of California at Irvine, 1983.

Gerson, Elihu M. "Career Contingencies in the Computing World." Unpublished paper, Pragmatica Systems, Inc., 1978.

Gibson, Cyrus F., and Richard L. Nolan. "Managing the Four Stages of EDP Growth." *Harvard Business Review*, January-February 1974.

Gilchrest, Bruce. "Wanted DP Professionals for the '80s." *Computerworld*, December 31, 1979.

————. "Technological Limitations on MIS Implementation." Paper presented at the ORSA/TIMS Conference, Colorado Springs, 1980.

Goetz, Martin A. "Unbundling: Will '80s Repeat the '60s?" *Computerworld*, April 14, 1980.

Gorry, G. Anthony, and Michael S. Scott Morton. "A Framework for Management Information Systems." *Sloan Management Review*, Fall 1971, pp. 55–70.

Gottmann, Jean. "Megalopolis and Antipolis: The Telephone and the Structure of the City." In Ithiel de Sola Pool, ed., *The Social Impact of the Telephone*. Cambridge, Mass.: MIT Press, 1977, pp. 303–317.

Gouldner, Alvin W. *Patterns of Industrial Bureaucracy*. New York: Free Press, 1954.

Greenbaum, Joan M. *In the Name of Efficiency: Management Theory and Shopfloor Practice in Data Processing Work*. Philadelphia: Temple University Press, 1979.

Gregory, Judith, and Karen Nussbaum. "Office Automation: A Threat to the

Job Satisfaction and Well-Being of Women Office Workers." *Proceedings of the 1981 Office Automation Conference,* March 1981, pp. 331–333.

"The Hacker Papers." *Psychology Today,* August 1980, pp. 62–69.

Hackman, J.R., and E.E. Lawler. "Employee Reactions to Job Characteristics." *Journal of Applied Psychology Monograph* 55 (1971):259–286.

Hackman, J.R., and G.R. Oldham. "Development of the Job Diagnostic Survey." *Journal of Applied Psychology* 60 (1975):159–170.

Handy, Charles. *Understanding Organizations.* London: Penguin, 1976.

———. "The Changing Shape of Work." *Organizational Dynamics,* Autumn 1980, pp. 26–34.

Hedberg, B.; A. Edstrom; W. Muller; and S.B. Wilpert. "The Impact of Computer Technology on Organizational Power Structures." In E. Grochla and N. Szyperski, eds., *Information Systems and Organization Structure.* Berlin: Walter de Gruyter, 1975, pp. 131–148.

Hedberg, B., and Enid Mumford, "Design of Computer Systems." In Louis E. Davis and James C. Taylor, eds., *Design of Jobs,* 2nd ed, Santa Monica, Calif.: Goodyear Publishing Co., 1979, pp. 44–53.

Hickson, D.J.; C.R. Hinings; C.A. Lee; R.E. Schneck; and J.M. Pennings. "A Strategic Contingencies' Theory of Intraorganizational Power." *Administrative Science Quarterly* 16 (1971):216–229.

Hinings, C.R.; D.J. Hickson; J.M. Pennings; and R.E. Schneck. "Structural Conditions of Intraorganizational Power." *Administrative Science Quarterly* 19 (1974):22–44.

Howard, Robert. "Brave New Workplace." *Working Papers for a New Society* 7(6) (1980):21–31.

Ives, Blake, and Margrethe H. Olson. "User Involvement in Information Systems: A Critical Review of the Empirical Literature." CRIS Working paper 15, GBA 81-07. New York: New York University, 1981.

James, Frank. "Get Vertigo Over Video Displays? Maybe It's a Case of Cyberphobia." *Wall Street Journal,* June 8, 1982.

Johnson, Bonnie, and Jerome Burke. "Learning to Learn: An Approach to Automating the Communication of Knowledge Workers." Paper presented at the Office Automation Conference, Houston, 1981.

Keen, Peter G.W. "Decision Support Systems." Unpublished working paper. Cambridge, Mass.: Massachusetts Institute of Technology, Center for Information Systems Research, 1979.

———. "Decision Support Systems and Managerial Productivity Analysis." Unpublished working paper. Cambridge: Massachusetts Institute of Technology, Center for Information Systems Research, 1980a.

———. "Information Systems and Organizational Change." Working paper

55. Cambridge, Mass.: Massachusetts Institute of Technology, Center for Information Systems Research, 1980b.

Keen, Peter G.W., and Gloria S. Bronsema. "Cognitive Style Research: A Perspective for Integration." Working paper 82. Cambridge: Massachusetts Institute of Technology, Center for Information Systems Research, 1981.

———. "Strategic Computer Education." Working paper 89. Cambridge, Mass.: Massachusetts Institute of Technology, Center for Information Systems Research, 1982.

Keen, Peter G.W., and Thomas J. Gambino. "Building a Decision Support System: The Mythical Man-Month Revisited." Working paper 57. Cambridge, Mass.: Massachusetts Institute of Technology, Center for Information Systems Research, 1980.

Kirchner, Jake. "DP Technology Creates More Jobs Than It Replaces, BLS Study Finds." *Computerworld*, November 2, 1981.

Kling, Rob. "The Impacts of Computing on the Work of Managers, Data Analysts and Clerks." Working paper 78-64. Irvine, Calif.: Public Policy Research Organization, 1978a.

———. "Information Systems as Social Resource in Policy-making." *Proceedings 1978 National ACM Conference* (1978b): 666–674.

———. "Defining the Boundaries of Computing on a Social Terrain." Working paper. Irvine, Cal.: University of California, Public Policy Research Organization, 1982.

Kling, Rob, and Elihu Gerson, "The Social Dynamics of Technical Innovation in the Computing World." *Symbolic Interaction* 1(1) (1977):132–146.

———. "Patterns of Segmentation and Intersection in the Computing World." *Symbolic Interaction* 1(2) (1978):24–43.

Kling, Rob, and Walt Scacchi. "Recurrent Dilemmas of Computer Use in Complex Organizations." *AFIPS Conference Proceedings* 48 (1979):107–115.

Kotter, John P. *Organizational Dynamics: Diagnosis and Intervention.* Reading, Massachusetts: Addison-Wesley Publishing Company, 1978.

Kraft, Philip. *Programmers and Managers: The Routinization of Computer Programming in the United States.* New York: Springer-Verlag, 1977.

———. "The Routinization of Computer Programming." *Sociology of Work and Occupations* 6(2) (1979):139–155.

Langefors, Borje. "Discussion of the Article by Bostrom and Heinen: MIS Problems and Failures: A Socio-Technical Perspective. Part 1: The Causes." *MIS Quarterly,* 2 (1978): 55–62.

Lattin, Don. "New Computers May Replace Many Controllers." *San Francisco Examiner,* August 9, 1981.

Laudon, Kenneth C. *Computers and Bureaucratic Reform.* New York: Wiley, 1974.

Lawler, Edward E., III, and John Grant Rhode. *Information and Control Systems in Organizations.* Santa Monica, Calif.: Goodyear Publishing, 1976.

Leavitt, Harold J., and Thomas L. Whisler. "Management in the 1980's." *Harvard Business Review,* November/December 1958, pp. 41–48.

Leduc, Nicole F. "Communicating Through Computers." *Telecommunications Policy,* September 1979, pp. 235–244.

Lehner, Urban C. "Japanese Firms Hire the Handicapped." *Wall Street Journal,* February 1, 1982.

Lippitt, Mary E.; J.P. Miller; and Jerry Halamaj. "Patterns of Use and Correlates of Adoption of an Electronic Mail System." *Proceedings of the American Institute of Decision Sciences,* Las Vegas, November 1980.

Locke, Benjamin. "Doctors and Computers: Why Are there Problems?" S.M. thesis, Massachusetts Institute of Technology, Sloan School of Management, 1980.

Lublin, Joann S. "Health Fears on VDTs Spur Union Action." *Wall Street Journal,* October 27, 1980.

Lucas, Henry C., Jr. *The Analysis, Design and Implementation of Information Systems.* New York: McGraw-Hill, 1981.

———. *Information Systems Concepts for Management,* 2nd ed. New York: McGraw-Hill, 1982.

Lundeberg, Mats; Goran Goldkukl; and Anders Nilsson. "Information Systems Development—A First Introduction to a Systematic Approach." Stockholm: ISAC, 1978.

———. *Information Systems Development: A Systematic Approach.* Englewood Cliffs, N.J.: Prentice-Hall, 1981.

Macrae, Norman. "The Coming Entrepreneurial Revolution: A Survey." *The Economist,* December 25, 1976, pp. 41–65.

Markus, M. Lynne. "Understanding Information System Use: A Theoretical Explanation." Ph.D. dissertation, Case Western Reserve University, 1979.

———. "Implementation Politics: Top Management Support and User Involvement." *Systems, Objectives, Solutions* 1 (1981):203–215.

———. "Power, Politics and MIS Implementation." *Communications of the ACM* 26(6) (1983a):430–444.

———. "Socio-Technical Systems: Concepts and Applications." In Terry Connolly, ed., *Scientists, Engineers and Organizations.* Monterey, Calif.: Brooks/Cole Publishing Company, 1983b, pp. 231–246.

Markus, M. Lynne, and Jeffrey Pfeffer. "Power and the Design of Accounting and Control Systems." *Accounting, Organizations and Society*, in press.

Martin, James. *Telematic Society*. Englewood Cliffs, N.J.: Prentice-Hall, 1981.

Mason, Richard O., and Ian I. Mitroff. "A Program for Research on Management Information Systems." *Management Science* 19(5) (1973):475–487.

Matlin, Gerald L. "IRM: How Will Top Management React?" *Infosystems*, October 1980.

Mattheis, Richard. "The New Back Office Focuses on Customer Service." *Harvard Business Review*, March/April 1979, pp. 146–159.

McFarlan, F. Warren; Robert L. Nolan; and David P. Norton. *Information Systems Administration*. New York: Holt, Rinehart & Winston, 1973.

McKenney, J.L., and Peter G.W. Keen. "How Managers' Minds Work." *Harvard Business Review*, May/June 1974.

Mechanic, David. "Sources of Power of Lower Participants in Complex Organizations." *Administrative Science Quarterly* 7 (1962):349–364.

"Meet Our Chief Designer, R2D2." *Wall Street Journal*, March 10, 1982.

Mertes, Louis H. "Doing Your Office Over—Electronically." *Harvard Business Review*, March/April 1981, pp. 128–135.

Meyer, Marshall W. "Automation and Bureaucratic Structure." *American Journal of Sociology* 74(3) (1968):256–264.

Mintzberg, Henry; Duru Raisinghani; and Andre Theoret. "The Structure of 'Unstructured' Decision Processes." *Administrative Science Quarterly* 21 (1976):246–275.

Mitroff, Ian I., and Richard Mason. "On Dialectical Pragmatism: A Progress Report on an Interdisciplinary Program of Research on the Dialectical Inquiry System." *Synthese* 47 (1981):29–42.

Mumford, Enid. *Systems Design for People*. Manchester: National Computing Center, 1971.

Mumford, Enid, and Mary Weir. *Computer Systems in Work Design—The ETHICS Method*. New York: Wiley, 1979.

"New Rivalry in Turnkey Systems." *Business Week*, June 23, 1980:107–108.

Noble, David F. "Social Choice in Machine Design: The Case of Automatically Controlled Machine Tools, and a Challenge for Labor." Extended version of article written for *Case Studies in the Labor Process*, Andrew Zimbalist, ed. New York: Monthly Review Press, 1979, pp. 18–50.

Nora, Simon, and Alain Minc. "L'Informatisation de la Societe." January 1978. Abstracted in *Chip Technology and the Labour Market*, P.M.H. Kendall; J. Crayston, et al. London: Metra Consulting Group, 1979, pp. 80–81.

Norsk Regnesentral (Norwegian Computing Center). "EDP og Sysselse (EDP and Employment)," March 1979. Abstracted in *Chip Technology and the Labour Market*, P.M.H. Kendall; J. Crayston, et al. London: Metra Consulting Group, 1979, p. 83.

"Note on the Minicomputer Industry in 1978." *Harvard Case Clearinghouse.* ICCH No. 9-579-177, 1979.

Nyhan, David. "Report from the Assembly Line: Many Jobs Are Lost Forever." *Boston Globe*, February 7, 1982.

Olson, Margrethe H. "Remote Office Work: Changing Work Patterns in Space and Time." *Communications of the ACM* 26(3) (1983):182–187.

Ouchi, William G., and Jerry B. Johnson. "Types of Organizational Control and Their Relationship to Emotional Well Being." *Administrative Sciences Quarterly* 23(Z) (1978):293–317.

Palme, Jacob. "Experience with the Use of the COM Computerized Conferencing System." Stockholm: Forsvarets Forskningsanstalt, 1981.

Pettigrew, Andrew M. "Information Control as a Power Resource." *Sociology* 6(2) (1972):187–204.

———. *The Politics of Organizational Decision-Making.* London: Tavistock, 1973.

Pfeffer, Jeffrey. *Organizational Design.* Arlington Heights, Ill.: AHM Publishing Corp., 1978.

———. *Power in Organizations.* Marshfield, Mass.: Pitman Publishing Inc., 1981.

Pfeffer, Jeffrey and Gerald R. Salancik. *The External Control of Organizations: A Resource Dependence Perspective.* New York: Harper & Row, 1978.

Phister, Montgomery. *Data Processing Technology and Economics*, 2nd ed. Santa Monica, Calif.: Digital Press, 1979.

Pomerantz, David. "Contract Programming." SM thesis, Massachusetts Institute of Technology, Sloan School of Management, 1981.

Poza, Ernesto, and M. Lynne Markus. "Success Story: The Team Approach to Work Restructuring." *Organizational Dynamics*, Winter 1980, pp. 2–25.

Robey, Daniel. "Information Technology and Organization Design." *University of Michigan Business Review* 28(5) (1976):17–22.

———. "Computers and Management Structure: Some Empirical Findings Re-Examined." *Human Relations* 30(11) (1977):963–976.

———. "Computer Information Systems and Organization Structure." *Communications of the ACM* 24(10) (1981):679–687.

Rockart, John F.; Christine V. Bullen; and Steve Leventer. "Centralization vs. Decentralization of Information Systems: A Framework for Decision Making." Working paper. Cambridge, Mass.: Massachusetts Institute of Technology, Center for Information Systems Research, 1977.

Rockart, John F., and Lauren S. Flannery. "The Management of End User Computing." Working paper 76. Cambridge, Mass.: Massachusetts Institute of Technology, Center for Information Systems Research, 1981.

Rockart, John F., and Steve Leventer. "Centralization vs. Decentralization of Information Systems: A Critical Survey of Current Literature." Working paper 23. Cambridge, Mass.: Massachusetts Institute of Technology, Center for Information Systems Research, 1976.

Schumacher, E.F. *Small Is Beautiful.* New York: Harper & Row, 1973.

Schuyter, Peter J. "Perils in Buying a Computer." *The New York Times,* September 11, 1980.

Sharpe, William. *The Economics of Computers.* New York: Columbia University Press, 1967.

Shirey, Robert W. "Management and Distributed Computing." *Computerworld,* October 20, 1980.

Shneiderman, Ben. "Human Factors Experiments in Designing Interactive Systems." *IEEE Computer* 12(12) (1979):9–19.

Shoor, Rita. "Travelers Says COBOL Conversion Could Cost a Cool $20 Million." *Computerworld,* March 23, 1981.

Sirbu, Marvin; Sandor Schoichet; Jay S. Kunin; Michael Hammer; and Juliet Sutherland. "OAM: An Office Analysis Methodology." Working paper. Cambridge, Mass.: Massachusetts Institute of Technology, Laboratory for Computer Science, 1981.

Stevens, Barry A. "Probing the DP Psyche." *Computerworld,* July 21, 1980.

Stewart, Rosemary. *How Computers Affect Management.* Cambridge, Mass.: MIT Press, 1971.

Strauss, George. "Tactics of Lateral Relationships." *Administrative Science Quarterly* 7(2) (1962):161–186.

"Systems that Slash Hospital Costs." *Business Week,* September 8, 1980:76A and 76E.

Taylor, James C. "Job Design Criteria Twenty Years Later." In Louis E. Davis and James C. Taylor, eds., *Design of Jobs,* 2nd ed., Santa Monica, Calif.: Goodyear Publishing Co., 1979, pp. 54–62.

———. "Quality of Working Life and White Collar Automation: A Socio-Technical Case." Working paper. Los Angeles, Calif.: UCLA Center for Quality of Working Life, 1981.

———. "Employee Participation in the Socio-Technical Systems Analysis of a Computer Operations Organization." *Journal of Occupational Behavior,* in press.

Toffler, Alvin. *The Third Wave*. Toronto: Bantam Books, 1980.

Turner, Jon A. "Achieving Consensus on Systems Requirements." *Systems, Objectives, Solutions* 1 (1981a):141–148.

————. "The Use of Organizational Models in Information Systems Research and Practice." Working paper GBA-81-21. New York: New York University, Graduate School of Business Administration, 1981b.

Uttal, Bro. "The Coming Struggle in Personal Computers." *Fortune*, June 29, 1981:84 v. 103.

Vitalari, Nichlas P. "The Evolution of Information Systems Support for Management." MISRC Working paper 78-01. Minneapolis: University of Minnesota, 1978.

Wagner, Harvey M. "A Managerial Focus on Systems Implementation." *Proceedings of a Conference Sponsored by the Center for Information Systems Research. The Implementation of Computer-Based Decision Aids*. Cambridge, Mass., April 1975.

Webster, Robin, and Leslie Miner. "Expert Systems: Programming Problem-Solving." *Technology*, January/February 1982, pp. 63–73.

Welke, Larry A. "The Software Product Business in the U.S.A.: Its Birth, Growth and Future for the Next Five to Ten Years." International Computer Programs, Inc., 1980.

Wetherbe, James C. "Evolution in Systems Analysis and Design." Paper presented at the Annual Conference of the Association for Systems Management, 1982.

Wildavsky, Aaron. "Policy Analysis Is What Information Systems Are Not." *Accounting, Organizations and Society*, 1978, pp. 77–88.

Withington, Frederick G. "The Organization of the Data Processing Function." In *Multicenter Networks*. New York: Wiley Business Data Processing Library, 1972, pp. 69–79.

Zuboff, Shoshannah. "New Worlds of Computer-Mediated Work." *Harvard Business Review*, September/October 1982, pp. 142–152.

Index